Chemistry with Computation

An Introduction to SPARTAN

Warren J. Hehre and
W. Wayne Huang

Wavefunction, Inc.
18401 Von Karman, Suite 370
Irvine, California 92715

ISBN 0-9643495-2-3

Printed in the United States of America

Table of Contents

1

SPARTAN: A Window into Computational Chemistry

The expanding role of calculations in chemistry and biochemistry and related fields, together with truly explosive advances in computer hardware and software, have prompted the development of *SPARTAN. SPARTAN* provides a convenient and familiar environment for chemists to carry out molecular mechanics calculations as well as semi-empirical and *ab initio* molecular orbital calculations and density functional calculations on organic, inorganic and organometallic molecules. It also facilitates calculations on collections of closely-related systems, as might be required to map conformation space, to "walk through a reaction coordinate", or to screen a set of compounds for a particular property or structural characteristic.

SPARTAN **is intended for chemists**, in particular, experimental chemists who, while they may have little or no background in theoretical methods, want to use calculations in much the same way as they now use experimental techniques such as NMR spectroscopy. Emphasis has been on providing a full range of "standard" computational tools, from molecular mechanics, to semi-empirical molecular orbital methods, to *ab initio* and density functional models, as well as facilitating interconnections among the available techniques. This reflects both a reliance on methods which are thoroughly documented and assessed, as well as the authors' view that no single method is likely to be ideal for every application.

Access to *SPARTAN* is by way of a graphical user interface which is highly functional yet rather simple and uncluttered. *SPARTAN* **looks like a Mac**, and chemists who are familiar with the Mac will have little difficulty finding their way around *SPARTAN.*

At the "front" of *SPARTAN's* interface are a series of molecule builders able to handle not only organic structures but also complex inorganic and organometallic systems as well as polypeptides. Simple tools have also been provided to construct transition states, to map reaction profiles and to explore conformation space. To facilitate comparisons among molecular systems, molecules may be optimally superimposed based either on their geometries or on some other quantity evaluated as a function of geometry.

The interface links the user to *SPARTAN's* "compute modules". These span a wide range of modern techniques from molecular mechanics (SYBYL, MM2 and MM3 force fields), to semi-empirical molecular orbital (MNDO, MNDO/d, AM1 and PM3 including newly developed parameterizations for transition metals), to *ab initio* (Hartree-Fock and MP2) and density functional schemes. *SPARTAN's* interface facilitates "mixing and matching" mechanics and quantum-chemical methods to tackle the problem at hand. Results from one level of calculation may easily be passed on to another higher level.

SPARTAN's interface provides a facile mechanism for making the output of molecular mechanics and quantum chemical calculations palatable to chemists. Not only the usual variety of structure models, but also molecular orbitals, electron and spin density distributions and molecular electrostatic and polarization potentials among other quantities, can be displayed, manipulated and animated in *real time*. As many quantities as desired can be displayed simultaneously, and visual comparisons easily made among them both for single molecules and among different molecules. We believe that SPARTAN's display capabilities provide chemists with a level of insight not previously available.

This brief guide is intended to serve as a practical introduction to SPARTAN, and more generally to the use of electronic structure calculations as a legitimate way of "doing chemistry". Four chapters follow. The largest is a "hands-on" tutorial illustrating the way in which calculations are carried out with SPARTAN. It touches on many of the more important of the program's functions, and demonstrates the interconnectivity of these functions in approaching problems. Next is a very brief and non-mathematical introduction to *ab initio* and semi-empirical molecular orbital methods and density functional methods, then a concise assessment of the performance of each of these methods, and finally a discussion of a number of practical aspects involved in actually doing calculations. With the exception of the tutorial which is intended to be used in conjunction with SPARTAN, that is, sitting down in front of the program, the other sections are independent of SPARTAN or for that matter any other program. Their focus is on methods which are generally applicable to chemical investigations and on strategies for effectively carrying out calculations.

None of the sections is comprehensive, and more complete accounts are available elsewhere. Specifically, **A SPARTAN Tutorial**[1] and **Experiments in Computational Organic Chemistry**[2] provide a more extensive series of practical examples covering all aspects of molecular mechanics and quantum chemical calculations, **Ab Initio Molecular Orbital Theory**[3] and **Critical Assessment of Modern Electronic Structure Methods**[4] survey the various classes of available methods and provide assessment of their performance, and **Practical Strategies for Electronic Structure Calculations**[5] focuses on a variety of practical concerns associated with carrying out calculations.

2

A SPARTAN Tutorial

This chapter comprises a series of simple exercises in the form of a self-guided tutorial, illustrating the way in which molecular mechanics and quantum chemical calculations may be set up and performed using SPARTAN, and the results of the calculations analyzed and interpreted. There has been no attempt either to illustrate SPARTAN's full functionality or to provide in-depth coverage of the functionality which is illustrated. Focus is on use of SPARTAN to calculate equilibrium and transition-state geometries, to search conformation space and to evaluate reaction energetics. **A** SPARTAN **Tutorial**[1] provides a more extensive set of practical examples covering a broader range of the program's features, and the SPARTAN **User's Guide**[6] describes in full the program's functionality. The exercises which have been provided are very simple, both for clarity and more practically to minimize computer time. In this regard, we have generally made use of semi-empirical models in preference to more costly *ab initio* and density functional models. The exercises do not need to be completed in order, nor is it necessary to wait until all computations in one exercise are finished before going on to the next exercise. This illustrates the intended mode of usage of SPARTAN, as an environment allowing job preparation and/or interpretation to proceed simultaneously with execution. The time required to complete all of the exercises should not be more than two to three hours.

To initiate SPARTAN, bring up a window on your workstation and type **spartan**. In a few seconds another window will appear on screen. Move it to the desired location and *click* with any mouse button. The main SPARTAN display appears as a blank screen except for a menu bar across the top.

The window may be moved about the screen by positioning the cursor in the area immediately above the menu bar, depressing either the left or the middle mouse button and moving the mouse. It may be scaled by *grabbing* one of the corners, depressing either the left or the middle mouse button and moving the mouse.

2.1

Building Molecules from Atomic Fragments

Among the simple building blocks incorporated into SPARTAN's *entry* builder are "atomic fragments". These constitute specification of atom type, e.g., carbon, and local environment, e.g., tetrahedral. A relatively few fragments allow construction of a wide variety of simple organic molecules.

Acetonitrile

We'll use acetonitrile to illustrate molecule construction from atomic fragments,

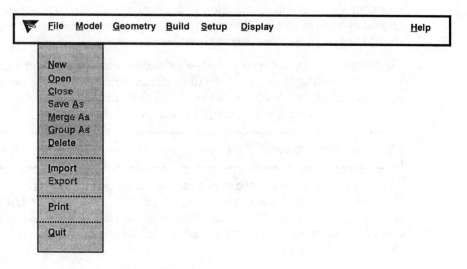

and then go on to outline the steps involved in setting up and submitting an *ab initio* molecular orbital calculation and following this, evaluation of the electric dipole moment and display of a graphical surface.

1. *Click* with the left mouse button on **File** from the menu bar,

▼	File	Model	Geometry	Build	Setup	Display		Help

New
Open
Close
Save As
Merge As
Group As
Delete

Import
Export

Print

Quit

and then *click* on **New** from the menu which appears. The *entry* builder screen which appears,

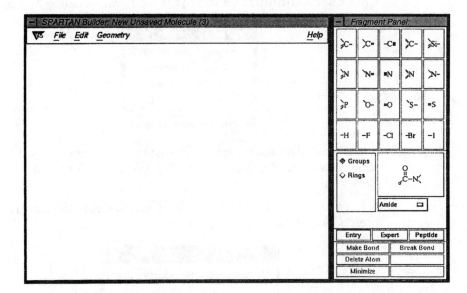

is divided into two parts, a *work area* on the left where molecules appear during the building process, and a *model kit* on the right which among other things contains a library of atomic fragments. *Click* (left mouse button) on tetrahedral, sp³ hybridized carbon from the library of fragments. The atom icon is now depressed indicating that it is *active*. Bring the cursor anywhere inside the *work area* and *click*. Tetrahedral carbon with its four unfilled valences indicated by "yellow vectors" will appear on screen.

> SPARTAN's builders connect atomic fragments (as well as groups, rings, ligands and chelates) through unfilled valences. Any remaining unfilled valences will automatically be converted to hydrogen atoms upon exiting.

2. Select (*click* on) linear, sp hybridized carbon from the *model kit*, and then *click* on the tip of one of the unfilled valences of the tetrahedral carbon atom in the *work area*. The linear carbon is automatically connected to the tetrahedral carbon by a single bond.

> SPARTAN's entry builder allows connection of atoms only through the same type of unfilled valence, e.g., single to single, double to double, etc.

3. Select linear, sp hybridized nitrogen from the *model kit*, and then *click* on the triple unfilled valence of the linear carbon in the *work area*. This connects carbon and nitrogen by a triple bond.

4. *Click* on the **Minimize** button in the *model kit*. A dialog box appears,

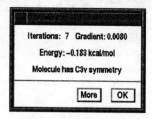

which gives the final molecular mechanics strain energy (-0.183 kcal/mol) and symmetry point group (C_{3v}). *Click* on **OK** to remove it from the screen.

5. Select **Quit** from the **File** menu (*click* on **File** from the menu bar and then *click* on **Quit**). A dialog appears,

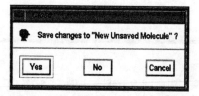

asking whether or not you wish to save your work. *Click* on **Yes**. Position the cursor inside the dialog which appears,

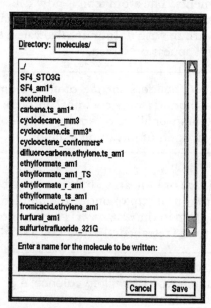

and type: "acetonitrile". *Click* on **Save** to exit the builder (or press the **Enter key**). Depending on the graphics device you are using, either a wire model or a tube model of acetonitrile will appear in the main screen.

wire model   tube model

This model can be manipulated (rotated, translated and zoomed) by using the mouse, if necessary, in conjunction with keyboard functions (see **Appendix B.4**). To rotate, drag the mouse while holding down the middle button; to rotate in the plane of the screen also hold down the **Shift key**. To translate the model, drag the mouse with the right button depressed. To zoom (translation perpendicular to the screen) hold down the **Shift key** in addition to the right button while dragging the mouse.

6. Select **Tube** (if a tube model was not already displayed), then **Ball and Wire**, then **Ball and Spoke** and finally **Space Filling** from the **Model** menu.

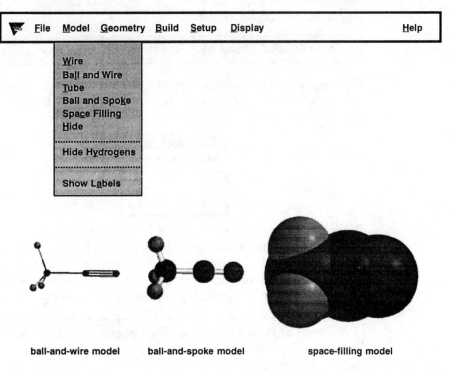

ball-and-wire model   ball-and-spoke model   space-filling model

7. Select **Ball and Wire** from the **Model** menu. This model, along with the wire model, are the only models for which atom numbers may be displayed. Select **Show Labels** from the **Model** menu. Numbers will appear next to the individual atoms together with atomic symbols. Remove the atom labels by selecting **Hide Labels** (which has replaced

Show Labels) from the **Model** menu.

8. Select **Distance** from the **Geometry** menu.

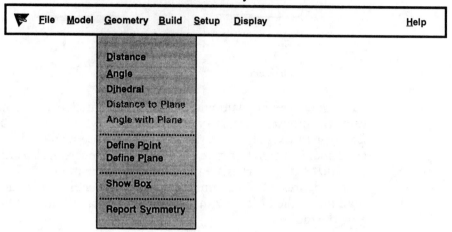

Whatever model was displayed will be replaced by a ball-and-wire model, and a message will appear in the menu bar requesting that two atoms (or a bond) be identified).

> Distance: Select 2 atoms, a bond or a distance constraint. (Enter "." to abort)

Clicking on the two atoms, e.g., the two carbons, will result in each being marked with a yellow sphere, and the distance between the two being displayed in a dialog box.

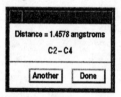

Where the two atoms are bonded, an alternative is to *click* on the bond. A miss on an atom or a bond is signaled by a bell; reposition the cursor and try again. Another distance may be obtained by selecting another pair of atoms (or another bond), and so forth. When you are done, *click* on **Done** inside the dialog box (or press the period "." key).

> Many of SPARTAN's functions may be terminated by the "." key.

The **Angle** and **Dihedral** entries from the **Geometry** menu operate in a similar manner. In the former case, the message displayed in the menu bar will be for three atoms (or two bonds) to be identified; in the latter case, four atoms (or three bonds) will need to be identified.

9. Select **Ab Initio** from the **Setup** menu.

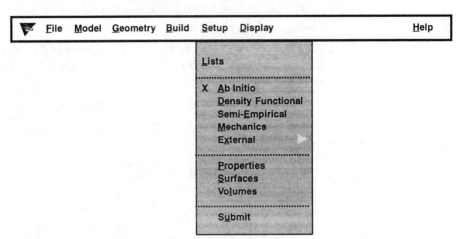

Perform the following operations (the order is unimportant) in the dialog which appears.

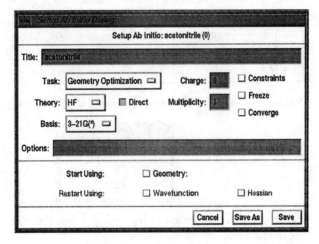

a. *Click* inside the **Title** box. Type "acetonitrile". The title is not used in the calculation, but will be reproduced in the text output. It is always a good idea to take the time and trouble to type in a title.

b. Select **Geometry Optimization** from the **Task** pull-down menu (*click* inside the text box to the right of **Task** and then *click* on **Geometry Optimization**). This specifies optimization of equilibrium geometry.

c. Select **HF** from the **Theory** pull-down menu. This specifies that Hartree-Fock theory is to be used.

d. Select **3-21G(*)** from the **Basis** pull-down menu. This specifies a calculation using the 3-21G split-valence basis set.

e. Verify that **Charge** is 0 and **Multiplicity** is 1. If you need to modify the contents of these boxes, *click* on the right side of the box, delete the contents (press the **Backspace key,** or move the cursor to the left with the left mouse button depressed), and then type the necessary value.

f. Turn off **Direct** if it is on (by *clicking* on the button to the left of **Direct).** This system is small and direct methods which use very little memory will not be necessary.

When you finish, *click* on **Save** to exit the dialog.

10. Select **Submit** from **Setup** menu. A dialog appears telling you that your calculation has been submitted.

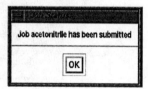

Click on **OK** to remove it from the screen.

After a SPARTAN calculation has been submitted menu items which could conceivably modify its input will not be accessible.

11. When the calculation has completed you will be notified.

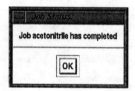

Click on **OK** to remove the dialog from the screen. Select **Output** from the **Display** menu.

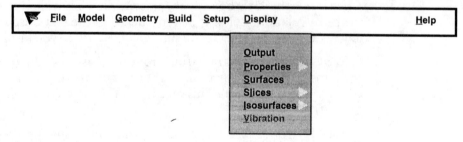

A dialog will appear on screen.

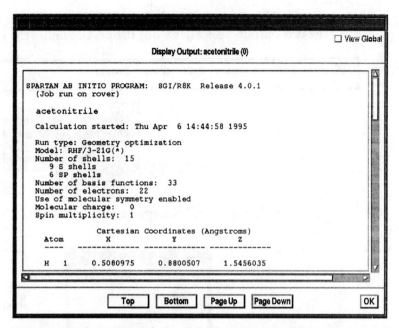

```
                                                              ☐ View Global
                        Display Output: acetonitrile (0)

SPARTAN AB INITIO PROGRAM:   SGI/R8K   Release 4.0.1
  (Job run on rover)

  acetonitrile

  Calculation started: Thu Apr  6 14:44:58 1995

  Run type: Geometry optimization
  Model: RHF/3-21G(*)
  Number of shells:  15
     9 S shells
     6 SP shells
  Number of basis functions:  33
  Number of electrons:  22
  Use of molecular symmetry enabled
  Molecular charge:    0
  Spin multiplicity:  1

                   Cartesian Coordinates (Angstroms)
      Atom         X            Y              Z
      ----     ------------- ------------- -------------

      H   1    0.5080975     0.8800507     1.5456035
```

[Top]   [Bottom]   [Page Up]   [Page Down]   [OK]

You can scan the output from the *ab initio* calculation by using the *scroll bar* at the right of the dialog, or by *clicking* on the **Top**, **Bottom**, **Page Up**, and **Page Down** buttons. The information at the "top" of the output dialog includes the title, task, basis set, number of electrons, charge, and multiplicity, as well as further details of the calculation. Below this is the initial geometry used for the calculation, the symmetry point group of the molecule that was maintained during the optimization (it should be C_{3v}), and the source of the trial Hessian (the Hessian is the matrix of second derivatives of the energy with respect to the Cartesian coordinates of the atoms).

Eventually, a series of lines appear, each beginning with "Cycle no:". These tell the history of the optimization process. Each line provides results for a particular geometry; "Energy =" gives the energy (in hartrees; 1 hartree = 627.5 kcal/mol) for this geometry, "rmsG =" gives the root-mean-square gradient, and "rmsD =" gives the root-mean-square displacement of atoms between cycles. Ideally, the energy will monotonically approach a minimum value for an optimized geometry, and rmsG and rmsD will each approach zero. If the geometry was not optimized satisfactorily an error message, such as "Optimization has exceeded N steps – Stop", will be displayed following the last optimization cycle. The user would then have been notified that the job had failed.

Near the end of the output are the final Cartesian coordinates, the final total energy (-131.191802 hartrees for acetonitrile with the 3-21G basis set), and the computation time. *Click* on **OK** to close the dialog.

The final total energy may be accessed without having to go through the text output. Select (*click* on) **Properties** under the **Display** menu,

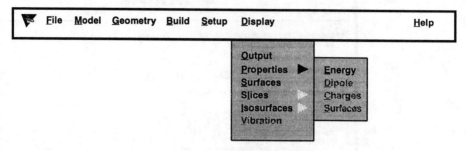

and then select (*click* on) **Energy** from the sub-menu which appears. A dialog box,

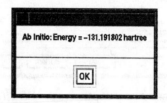

provides the energy. *Click* on **OK** to remove the dialog.

12. Select **Properties** from the **Setup** menu; the following dialog appears.

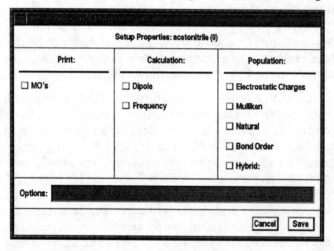

Click on the box to the left of **Dipole** (to request calculation of the electric dipole moment) and then *click* on **Save** to close the dialog. Again select **Submit** from the **Setup** menu, and handle the various message boxes as you did during the geometry optimization.

When the dipole moment calculation has completed, select **Properties** from the **Display** menu and then **Dipole** from the sub-menu which results. A dialog box,

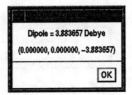

Dipole = 3.883657 Debye

(0.000000, 0.000000, –3.883657)

OK

gives the total dipole moment (3.88 debyes) as well as its Cartesian components. In addition, the dipole moment vector (+ +———→ -) is displayed on screen attached to the moleular model. *Click* on **OK** to remove both the dialog and the vector.

13. Select **Surfaces** from the **Setup** menu; a dialog containing three pull-down menus appears.

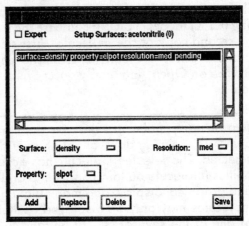

☐ Expert   Setup Surfaces: acetonitrile (0)

surface=density property=elpot resolution=med pending

Surface: density   Resolution: med

Property: elpot

Add   Replace   Delete   Save

Select **density** from the **Surface** menu (*click* inside the text box to the right of **Surface** and then *click* on **density**), **elpot** from the **Property** menu and **med** from the **Resolution** menu, respectively. This will specify calculation of a total electron density surface onto which the value of the electrostatic potential has been mapped. *Click* on **Add**; the line "surface=density property=elpot resolution=med pending" appears in the large text box at the top of the dialog (as shown above). If you make a mistake, *click* on the line describing the unwanted surface (it will then be highlighted), and then *click* on **Delete**. *Click* on **Save** to close the dialog.

14. Submit the job (**Submit** from the **Setup** menu), and handle the various message boxes as you did during the geometry optimization. When the calculation has completed, select **Surfaces** from the **Display** menu.

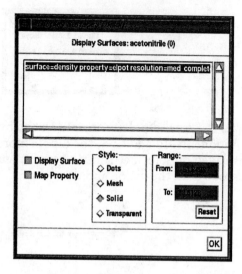

Click on the text string at the top of the box inside the window to highlight it. *Click* on the button to the left of **Display Surface** and then select **Solid** under **Style**. *Click* on **OK** to exit the dialog. Note the striking resemblance of the density surface to a conventional space-filling (CPK) model. You can compare the two directly by first opening a second copy of the molecule. First select **Open** from the **File** menu, then select "acetonitrile" from the file browser which appears, then *click* on **Open**, and finally select **Space Filling** from the **Model** menu.

> Two or more molecules may be simultaneously displayed in SPARTAN's main window, although only one molecule may be selected. The selected molecule has access to all menu capabilities (molecule building, job setup and submission, and text and graphical display, and manipulation), while non-selected molecules may only be displayed as static images. Selection of one of the several sets of images currently on screen occurs by moving the cursor anywhere onto the image and *clicking* with the left mouse button.

A more revealing comparison follows by displaying a transparent density surface along with a space-filling model, i.e., on the same structure. *Click* on the density surface (this selects it as the active molecule), reenter the **Surfaces** dialog under the **Display** menu, highlight (*click* on) the text string and then select **Transparent** under **Style**. Exit the dialog by *clicking* on **OK**. The image is now transparent and you can clearly see the model underneath.

Transparent objects are poorly displayed on low-resolution graphics devices. In this case, you can either use a mesh display instead of a transparent display (select **Mesh** under **Style**) or press the "1" key to produce a higher-resolution static transparent image. In the latter instance, any movement will return the display to low-resolution mode.

Replace whatever model you are using by a space-filling model (**Space Filling** from the **Model** menu). You can now see that the two images (electron density surface and space-filling model) are nearly identical.

15. Return to a tube model (**Tube** from the **Model** menu), and then reenter the **Surfaces** dialog (**Display** menu) one last time. First highlight the text string, then select **Solid** under **Style**, and finally *click* on the button to the left of **Map Property.** This requests that the electrostatic potential (the energy of interaction between a point positive charge and the nuclei and electrons of the molecule) be color encoded on top of the density surface. Exit the dialog by *clicking* on **OK.** Examine the graphic which appears on screen. The shape of the surface is the same as before, and the colors indicate values of the electrostatic potential evaluated on this surface; colors near red correspond to negative potential (stabilizing interaction between the molecule and the positive test charge), while colors near blue correspond to positive potential. The nitrogen will appear as an electron-rich site (red), and the hydrogens will appear as electron-poor sites (blue). This is reasonable as nitrogen is the most electronegative atom in the molecule and the hydrogens are the most electropositive. This graphic suggests that the dipole moment which you calculated earlier has the hydrogens at the positive end and the nitrogen at the negative end. You might go back and examine the dipole moment vector again (select **Properties** under the **Display** menu and then **Dipole** under the sub menu which results) to see if this is indeed the case.

16. Remove both models of acetonitrile from the screen by selecting each in turn and then selecting **Close** from the **File** menu.

2.2

Building Molecules from Functional Groups and Rings

Organic chemistry is organized around functional groups, groups of atoms the structure and properties of which are roughly the same in every molecule. *SPARTAN* simplifies the construction of organic molecules which contain one or more functional groups by providing a small library of predefined structures, where the functional groups included in the library can easily be modified. For example, the ester group may be used to build, among other things, a carboxylic acid (no changes required), a carboxylate anion (by deleting an unfilled valence from oxygen), or an ester (by adding tetrahedral carbon to the unfilled valence at oxygen), i.e.,

carboxylic acid   carboxylate anion   carboxylic acid ester

Polyatomic rings are also common components in organic molecules, and SPARTAN simplifies their incorporation by providing a library of commonly-encountered hydrocarbon rings. While only hydrocarbon rings are included in the library, these can easily be modified by atom replacement. For example, pyridine can be built from benzene by selecting aromatic nitrogen from the list of atomic fragments, and then *double clicking* (two *clicks* in rapid succession) on one of the carbons.

select
and *double click* on carbon

benzene   pyridine

Functional groups may also be modified in this manner, and SPARTAN's *entry* builder performs atom replacement subject to usual valence rules.

Furfural

We'll construct furfural,

using both rings and functional groups, and then use it to illustrate another of SPARTAN's graphics display modes.

1. Enter the builder (**New** from the **File** menu). *Click* on **Rings** in the *model kit* and then *click* inside the text box immediately below the icon of whatever ring is presently displayed. The following menu appears.

Rings:

Cyclopropyl
Cyclobutyl
Cyclopentyl
Cyclohexyl
....
Phenyl
Naphthyl
Anthryl
Phenanthryl

Click on **Cyclopentyl**, move the cursor anywhere in the *work area* and *click*. Cyclopentane appears on screen.

2. Select sp^3 oxygen from the atomic fragments and then *double click* on one of the carbons on the cyclopentyl ring. An oxygen will be substituted for one of the methylene groups leaving you with tetrahydrofuran.

3. Introduce double bonds into the ring by *clicking* on **Make Bond*** in the *model kit*. Next, *click* on one of the unfilled valences of a carbon bonded to the oxygen, then *click* on one of the unfilled valences of the adjacent carbon. A double bond will appear between the two carbons. In a similar way, introduce a second double bond to make furan. Don't worry that the molecule is highly distorted.

4. Add the carbonyl group to the ring by first *clicking* on **Groups** from the *model kit*, then *clicking* inside the text box immediately below the icon of whatever functional group is presently displayed,

Groups:

Alkynyl
Allenyl
Amide
Carbonyl
Cyano
Ester
Nitro
Nitroso
Phosphinyl
Sulfonyl
Sulfoxide
Vinyl

then *clicking* on **Carbonyl**, and finally *clicking* on the appropriate unfilled valence on furan.

5. *Click* on **Minimize** to produce a refined structure. It should have C_S symmetry. Exit the builder (**Quit** from the **File** menu) and supply the name: "furfural_am1".

6. Select **Semi-Empirical** from the **Setup** menu.

---

* **Make Bond** may also be accessed under the **Edit** menu in the builder.

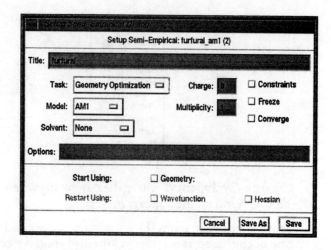

Provide a title and specify **Geometry Optimization** under **Task** and **AM1** under **Model** (**None** under **Solvent**). The default **Charge** (0) and **Multiplicity** (1) values are correct. *Click* on **Save** to exit the dialog.

7. Let's calculate and display the HOMO of furfural in two different ways, both from a series of HOMO values inside a box enclosing the molecule. Select **Show Box** from the **Geometry** menu. This displays a wire frame around the molecule which will be used to delineate the extent of the volume of graphical data to be calculated.

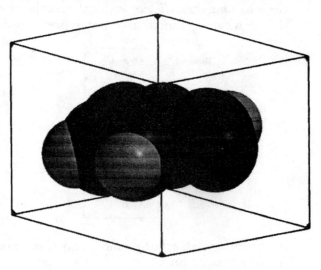

We want this volume to enclose the van der Waals surface, so replace whatever model is presently displayed by a space-filling model (**Space Filling** from the **Model** menu). The box can be expanded or contracted by positioning the cursor onto any of its exposed corners, depressing the left mouse button and moving the mouse. The most exposed face (together with the associated parallel face) will expand and contract (the white frame around the two windows which move will be re-placed by a red frame). Size the box such that the space-filling model fits comfortably inside. When you are done, remove the frame by se-lecting **Hide Box** from the **Geometry** menu.

8. Return to a tube model (**Tube** under the **Model** menu). Select **Volumes** under the **Setup** menu.

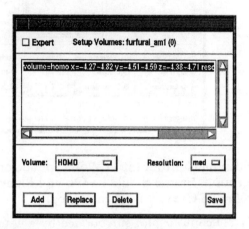

Select **HOMO** from **Volume** and **med** from **Resolution**. *Click* on **Add**. This results in the text line shown in the dialog above and will instruct *SPARTAN* to produce a volume of highest-occupied molecular orbital values which fill up the box. Exit the dialog by *click*ing on **Save**.

9. Submit the job. When completed, *click* on the **Display** menu, then *click* on **Slices**,

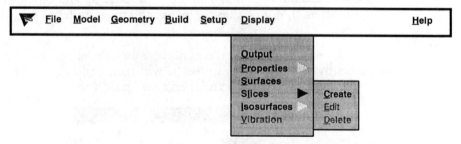

and finally *click* on **Create** from the sub menu which results.

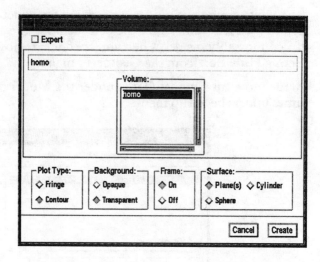

Click on **Contour** under **Plot Type** and on **Transparent** under **Background**. **Frame** should be **On** and **Plane(s)** selected under **Surface**. *Click* on **Create** to exit the dialog.

10. A 2D contour plot surrounded by a white frame will be displayed. *Click* on the frame (it will become yellow) to make the contour the selected graphical object; it can now be manipulated (translated and rotated) independently of the molecule like any other graphical object. The contour can be resized by holding down both the right mouse button and the **Shift key** while dragging the mouse. *Click* on the structure model; it will now be the selected object and manipulations refer to the total graphic. Manipulating both the plane alone and the total graphic, try to position the slice parallel to, but above or below the plane of the molecule. As you move the slice up and down (through the molecule), notice that the phase and magnitudes of the HOMO change.

11. A scale may be added to the graphic. Select the contour as the active object by *clicking* on its frame (it will turn yellow), *click* on **Slices** under the **Display** menu and then *click* on **Edit**.

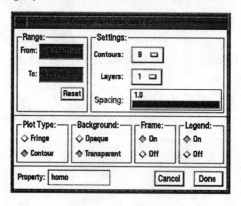

Click on **On** under **Legend** and exit the dialog by *clicking* on **Done**. A scale will appear which may be moved about the screen. *Click* on it to designate it as the active graphic (a yellow box will then enclose it) and move the mouse with the right button depressed. The scale may also be "zoomed" by simultaneously depressing the right button and the **Shift key** while moving the mouse.

12. Remove the contour plot (and its associated scale) from the screen by selecting it (*click* on the white frame; it will turn yellow), and then selecting **Delete** under **Slices** under the **Display** menu.

13. Volume data may also be used to construct isosurfaces, that is, surfaces of constant value. The value of the isosurface does not need to be specified prior to calculation, but rather can be adjusted "in real time" following calculation of the volume data. *Click* on **Isosurfaces** under the **Display** menu and then *click* on **Create** in the sub-menu which results.

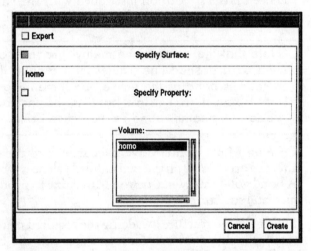

Only a single quantity (the HOMO) is available and the only action you need to take is to *click* on **Create**. The dialog disappears and in its place a HOMO isosurface appears. An isosurface is "attached" to the structural model and the two cannot be moved independently.

14. Remove all images from screen (**Close** from the **File** menu).

Camphor

Construction of camphor,

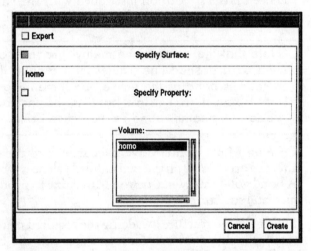

allows us to further illustrate builder functions. The molecule also provides an interesting example of application of graphical modeling techniques to the investigation of stereoselectivity.

1. Enter the builder (**New** from the **File** menu). *Click* on the **Rings** button, *click* inside the text box which appears to the right and immediately below whichever ring icon is presently displayed, and then *click* on **Cyclohexyl**. Position the cursor anywhere in the *work area* and *click*. Cyclohexane will appear. Select the sp^2 carbon from the *model kit*. *Double click* on any carbon. One of the sp^3 carbons on cyclohexane will be replaced by an sp^2-hybridized carbon. Select sp^2 oxygen from the *model kit*. Position the cursor on the double unfilled valence on the sp^2 carbon and *click*. You have made cyclohexanone. Select sp^3 carbon from the *model kit*. *Click* on an *axial* free valence on carbon adjacent to the carbonyl carbon. While holding down the middle mouse button and simultaneously the **space bar**, move the mouse to align an empty valence on the sp^3 carbon just added with the empty *(equatorial)* valence on the opposite side of the six-membered ring.

> Note that the bond to the methyl group is connected to the ring by a dashed line (rather than a solid line). This designates this bond as the active bond, meaning that we can rotate about it. The active bond is the last bond formed. To designate another bond as the active bond, *click* on it.

Click on **Make Bond**. Move the cursor into the *work area* and, one after the other, *click* on the two unfilled valences which you just aligned. A bond will be drawn. *Click* on **Minimize** to produce a refined (intermediate) structure.

Finish off the structure by adding the required methyl groups (three in total). Finally, *click* on **Minimize** and exit the builder by selecting **Quit** from the **File** menu. Name the molecule: "camphor_am1".

2. Select **Semi-Empirical** from the **Setup** menu. Provide a title and set the entries under **Task**, **Model** and **Solvent** to **Geometry Optimization**, **AM1**, and **None**, respectively. The default values for **Charge** (0) and **Multiplicity** (1) are corrrect. Exit the dialog by *clicking* on **Save**. Submit the job (**Submit** from the **Setup** menu). At any time during the geometry optimization, the job may be examined by selecting **Output** under the **Display** menu. In fact, you can "follow the job" by going to the bottom of the output dialog (*click* on **Bottom** inside the dialog). When it has finished, proceed to the next step.

3. Camphor undergoes nucleophilic attack at the carbonyl carbon. We would expect that the molecule's lowest-unoccupied molecular orbital (the LUMO) to be localized on the carbonyl group. To visualize the LUMO, enter the **Surfaces** dialog (**Setup** menu) and select **LUMO** from the **Surfaces** menu (**none** from the **Property** menu and **med** from the **Resolution** menu). *Click* on **Add**. Also request calculation of a surface of total electron density onto which the absolute value of the

LUMO has been encoded in color. Select **density** from **Surfaces** and **LUMO** from **Properties** (**med** from **Resolution**). *Click* on **Add**. Exit the dialog by *clicking* on **Save**, and resubmit the job (**Submit** under the **Setup** menu).

4. After the graphics calculations have completed, select **Surfaces** under the **Display** menu. *Click* on the line "surface = lumo, ..." in the text box, *click* on the button to the left of **Display Surface**, select **Solid** under **Style** and finally *click* on **OK**. The dialog box disappears and a solid model depicting the LUMO of camphor appears. As expected, this is a π^* orbital primarily localized on the carbonyl group.

5. Reenter the **Surfaces** dialog (**Display** menu). Turn off the display of the LUMO by *clicking* on the line "surface = lumo, ..." in the text box and then on the button to the left of **Display Surface**. Display the density surface as a mesh by *clicking* on the line, "surface=density..." in the text box, *clicking* inside the box to the left of **Display Surface**, and then selecting **Mesh** under **Style**. *Click* on **OK** to exit the dialog. The resulting surface allows the skeletal model of camphor to be seen below. Select **Space-Filling** under the **Model** menu. Note the overall similarity of the space-filling model and the electron density surface. This is not coincidental; the value of the electron density used to define the surface has been chosen to provide an overall size similar to that which would result from use of van der Waals contact radii, i.e., the same radii as used in the definition of CPK models.

6. Return to a tube model (**Tube** under the **Model** menu) and reenter the **Surfaces** dialog (**Display** menu). *Click* on the line "surface=density...", select **Solid** under **Style** and then *click* on the button to the left of **Map Property**. *Click* on **OK** to exit the dialog. The surface which now appears corresponds to a van der Waals contact surface (see above) onto which the absolute value of the lowest-unoccupied molecular orbital has been superimposed. Colors near red indicate small values of the LUMO (near zero), while colors near blue indicate large values of the LUMO. Note the "blue spot" directly over the carbonyl carbon. This corresponds to the maximum value of the LUMO and is where nucleophilic attack will occur. This result is, of course, not surprising.

7. Turning the molecule, you will see that the "blue spot" over the *axial* face of the carbonyl carbon is more intense than over the *equatorial* face. This indicates preferential attack by nucleophiles onto the *axial* face, in accord with the usual observation. We can quantify this difference by measuring the value of the LUMO on these two surfaces. Select **Properties** under the **Display** menu, and then select **Surfaces** from the sub-menu which appears. A message will appear at the top of the screen,

> Properties: Select on surface. (Enter "." to abort)

instructing you to *click* on any portion of the displayed surface. Turn the surface such that you can clearly see the *axial* face of the carbonyl carbon, and *click* on the area of maximum blue. A copy of the cursor remains on the surface and a dialog appears on the screen.

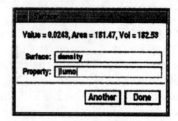

This provides the (absolute) value of the LUMO at the surface point you have selected. Note the value, and then turn the object over such that you can now see the *equatorial* face of the carbonyl carbon, and *click* on the point of maximum blue on this face. Do the surface values support your qualitative conclusions from viewing the image?

8. *Click* on **Done** to remove the dialog. Remove the model and the surface from the screen using **Close** from the **File** menu.

2.3

Building Inorganic and Organometallic Molecules

The majority of organic molecules are made up of a relatively few elements and obey conventional valence rules. They may be easily constructed using *SPARTAN's entry* builder. There are, however, numerous molecules which incorporate other elements, or which do not conform to normal valence rules, or which involve ligands or chelating groups. These need to be constructed using *SPARTAN's expert* builder.

Sulfur Tetrafluoride

A simple hypervalent molecule such as sulfur tetrafluoride,

cannot be constructed using *SPARTAN's entry* builder. This is because sulfur is not in its normal bent dicoordinate geometry, but rather in a trigonal bipyramid geometry with one of the *equatorial* positions vacant. SF_4 is, however, easily made with the *expert* builder screen. We'll also use this example to illustrate calculation and display of atomic charges.

1. Enter the *expert* builder by going into the *entry* builder (**New** from the **File** menu) and then *clicking* on **Expert** in the *model kit*. The screen which appears,

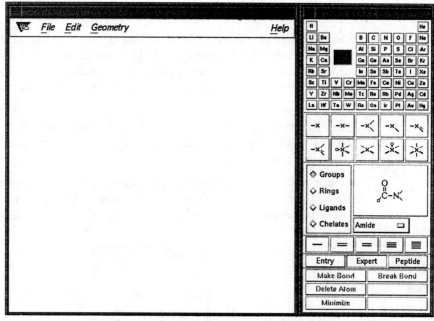

comprises a *Periodic Table* with main-group elements at the top, and transition metals at the bottom. Other elements can be specified by typing their atomic symbols into the text box at the center of the *Periodic Table*. Immediately below the *Periodic Table* is a listing of "atomic hybrids". Further down the *model kit* are buttons marked **Rings**, **Groups**, **Ligands** and **Chelates**, the first two of which are the same as found in the *entry* builder. Next, there is a listing of "bond types": single, partial double, double, triple and quadruple. The default type is single, and selection of one of the other types results in scaling of standard bond distances kept in the builder's library. Buttons at the bottom of the *model kit* and menus above the *work area* control other builder functions, and are the same as those for the *entry* builder screen.

2. *Click* on S in the *Periodic Table* and the five coordinate trigonal bipyramid structure from the list of atomic hybrids. Position the cursor inside the *work area* and *click*. A trigonal bipyramid sulfur will appear on screen.

3. *Click* on **F** in the *Periodic Table* and the one-coordinate entry from the list of atomic hybrids. One after the other, *click* on both *axial* free valences of sulfur, and two of the three *equatorial* free valences.

4. It is necessary to delete the remaining free valence (on an *equatorial* position); otherwise it will become a hydrogen upon exiting the builder. *Click* on the **Delete Atom*** button in the *model kit* and then *click* on the remaining *equatorial* free valence.

5. *Click* on **Minimize**. Molecular mechanics optimization will result in a structure with C_{2v} symmetry. *Click* on **OK** to remove the dialog, select

---

*Delete Atom may also be accessed under the **Edit** menu in the builder.

Quit from the **File** menu to exit the builder, and supply the name: "sulfurtetrafluoride_321G".

6. Select **Ab Initio** from the **Setup** menu. Type in a title, and specify a single-point energy Hartree-Fock (HF) calculation at the 3-21G$^{(*)}$ level, and make certain that **Direct** is selected. The default values for **Charge** (0) and **Multiplicity** (1) are correct. *Click* on **Save** to exit the dialog.

7. Select **Properties** from the **Setup** menu. *Click* inside the box to the left of **Electrostatic Charges** under **Population** to specify that atomic charges will be obtained from fits to electrostatic potentials. *Click* on **Save** to exit the dialog.

8. Submit the job (**Submit** under the **Setup** menu). When completed, select **Charge** from the **Properties** sub-menu (**Display** menu). You will be instructed to *click* on an atom, e.g., sulfur, after which a dialog will appear.

To obtain the charge on another atom simply *click* on it. Inspect all the atomic charges on SF$_4$ (by *clicking* on the appropriate atoms). Note that sulfur carries a large positive charge. When you are finished, *click* on **Done**.

9. The usual picture is that the remaining *equatorial* position in sulfur tetrafluoride holds a non-bonded pair of electrons (a "lone pair"). In the case of a 3-21G$^{(*)}$ calculation, the lone pair is most closely represented by the highest occupied molecular orbital. Let's plot the HOMO and see to what extent this simple picture is accurate.

   Enter the **Volumes** dialog under the **Setup** menu. Specify **HOMO** from the **Volumes** menu and **med** from the **Resolution** menu. *Click* on **Add**, and then *click* on **Save** to exit the dialog.

10. Submit the job (**Submit** under the **Setup** menu). When the calculation has completed, select **Isosurfaces** under the **Display** menu and then **Create** under the sub-menu which follows. "homo" is the only entry inside the dialog which appears, and all you need to do to display it is to *click* on the **Create** button. Does the image you see clearly identify the orbital as corresponding to a lone pair on sulfur?

11 to 13 optional

Let's perform a single-point local density functional calculation to see if there are qualitative differences in charges and description of the highest-occupied molecular orbital from those given by the 3-21G$^{(*)}$ *ab initio* calculation.

11. Make a copy of: "sulfurtetrafluoride_321G" (**Save As** from the **File** menu); name it: "sulfurtetrafluoride_vwn.dnp"

12. Select **Density Functional** from the **Setup** menu.

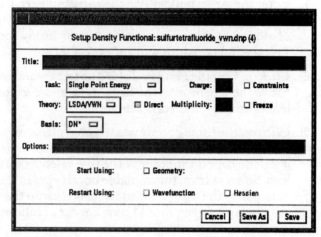

Once inside the dialog, specify **Single Point Energy** for **Task**, **LSDA/VWN** for **Theory** and **DN*** for **Basis**. Make certain that **Direct** is selected. The default **Charge** (0) and **Multiplicity** (1) are correct. Exit the dialog by *clicking* on **Save**. You will be presented with a message indicating that previous information (from the 3-21G$^{(*)}$ *ab initio* calculation) is to be overwritten. *Click* on **Save**.

The previous selections which you made in the **Properties** dialog (calculation of electrostatic charges) and in the **Volumes** dialog (calculation of the HOMO) are still in place. You can check this if you wish by entering the two dialogs one after the other.

13. Submit the job. When it is done (it will take a few minutes), examine the calculated charges (see 8 above) and the HOMO isosurface (see 10 above). Do you see any qualitative differences in charges or structure of the HOMO in going from the *ab initio* to the density functional calculation?

14. Remove the structural model and graphics from the screen (**Close** from the **File** menu).

Dicyclopentadienyl, Methyl Tantalum Methylidene

This simple transition metal alkylidene complex,

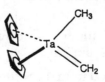

illustrates additional features of the *expert* builder, as well as provides an example of the use of the PM3 (tm) semi-empirical model for transition metals.

1. Enter the *expert* builder (via the *entry* builder). *Click* on **Ta** in the *Periodic Table* and the four-coordinate tetrahedral structure from the list of atomic hybrids. Position the cursor inside the *work area* and *click*. Tetrahedral tantalum will appear on screen.

2. *Click* on **C** in the *Periodic Table* (the tetrahedral hybrid is still selected). Move the cursor into the *work area,* and *click* on one of the free valences on the tetrahedral tantalum.

3. *Click* on the three-coordinate trigonal planar structure from the list of atomic hybrids (carbon is still selected). Move the cursor into the *work area,* and *click* on another of the free valences on the tetrahedral tantalum. Select **=** from the available bond types in the *model kit,* move the cursor into the *work area,* and *double click* on the Ta-C (sp²) bond. The structure on screen should now appear as follows:

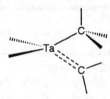

4. *Click* on the **Ligands** button in the *model kit, click* inside the text box immediately below whichever ligand is presently displayed and *click* on **Cyclopentadienyl.**

Ligands:

Acetylene
Ethylene
Allyl
Butadiene
Cyclopentadienyl
Benzene
Carbon Monoxide
Nitrogen Oxide
Ammonia
Water
Phosphine

Move the cursor into the *work area,* and then one after the other *click* on the two remaining free valences on tantalum. Two cyclopentadienes are added as ligands to the metal.

5. *Click* on **Minimize.** After the molecular mechanics optimization has completed, select **Quit** from the **File** menu and supply the name: "Cp2TaCH3 = CH2_pm3tm".

6. Select **Semi-Empirical** from the **Setup** menu. Type in a title, and specify **Geometry Optimization** under **Task** and **PM3 (tm)** under **Model** (**None** under **Solvent**). The default **Charge** (0) and **Multiplicity** (1)

values are correct. *Click* on **Save** to exit the dialog.

7. Submit the job. When the optimization has completed (it will take a few minutes), examine the final methyl-tantalum and methylidene-tantalum carbon-carbon bond distances (**Distance** under the **Geometry** menu). For comparison, the experimental values are 2.25 and 2.03Å, respectively.

8. Remove the structural model from the screen (**Close** from the **File** menu).

2.4

Flexible Molecules

One of the greatest difficulties encountered in applications of computation is the assignment of conformation around single bonds. Clearly, detailed conformation dictates overall molecular size and shape, influences molecular properties and controls chemical reactivity and selectivity. SPARTAN is able to search conformation space both for acyclic systems, as well as for molecules incorporating flexible rings. Semi-empirical, *ab initio* and density functional molecular orbital models may be used, as well as molecular mechanics techniques. Conformation searching is, however, very intensive computationally, and more often than not we will need to be satisfied using complete surveys carried out with molecular mechanics or at best semi-empirical techniques, followed by limited higher-level calculations to better establish relative conformer populations. SPARTAN's list processing abilities are ideally suited for this kind of approach.

Cyclooctene

Small and medium ring cycloalkenes adopt *cis* arrangements about the incorporated double bond.

cis   *trans*

Trans isomers of cyclooctene and smaller cycloalkenes are very high in energy (relative to *cis* isomers), and are difficult to isolate, while *trans* isomers of larger systems should be more accessible energetically.

In order to obtain the difference in energy between *cis* and *trans* isomers for any given cycloalkene, it is necessary to find the lowest-energy conformation (*global minimum*) of each. We illustrate this for cyclooctene using the MM3 molecular mechanics model for identifying possible conformations, and later the semi-empirical AM1 model to assess relative energies of stable forms. In so doing, we show how the results of a conformation search performed at one level of calculation may easily be passed on for further analysis using another level of calculation.

1. Go into the *entry* builder and construct *cis*-cyclooctene from a series of sp³ and sp² hybridized carbons. Don't worry about its conformation. Minimize and exit the builder supplying the name: "cyclooctene.cis_mm3".

2. Select **Conformer Search** from the **Build** menu.

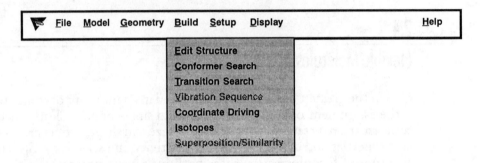

In the dialog which appears,

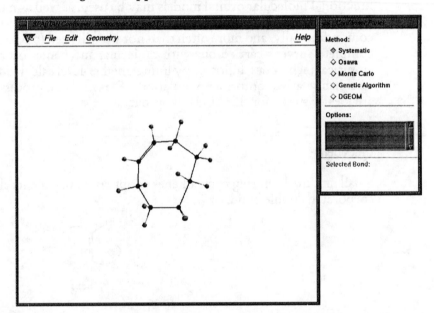

click on the button to the left of **Osawa** under **Method**. This is presently the only searching technique available for rings within *Spartan*. Next, position the cursor on top of any one of the single bonds incorporated into the ring and *double click*. A small gold cylinder will appear around the selected bond. This indicates that any ring which incorporates this bond will be searched for the lowest-energy conformation. Exit the dialog by selecting **Quit** from the **File** menu.

3. Select **Mechanics** from the **Setup** menu.

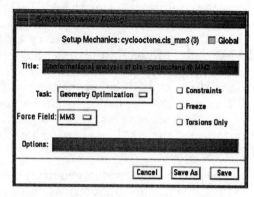

Type in a title, select **Geometry Optimization** under **Task** and **MM3** under **Force Field**. Exit the dialog by *clicking* on **Save**.

4. What you have done in steps 2 and 3 above is to request that a conformation search is to be carried out with full geometry optimization and using the MM3 force field. Submit the job. When the calculation has completed you will be notified that a list (comprising the conformers of cyclooctene) is to be formed.

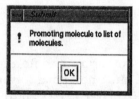

Click on **OK**. In a few seconds a new dialog (the **Lists** dialog) will appear.

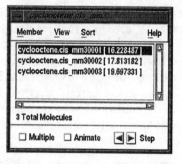

This allows inspection of the individual conformers inside the list, and will be present whenever a list member is active. Select **Show Energy** under the **View** menu in this dialog to display the MM3 strain energies. This initial set of conformers is already sorted from lowest energy (top) to highest energy (bottom). You can move between conformers by *clicking* on the entry in the large text box contained in this dialog.

5. Recalculate the structures and relative energies of the full set of cyclooctene conformers using the AM1 semi-empirical model. This can be done in a single step. Enter the **Semi-Empirical** dialog (under the **Setup** menu). Select **Geometry Optimization** under **Task** and **AM1** under **Model** (**None** under **Solvent**). The default **Charge** (**0**) and **Multiplicity** (**1**) values are correct. Make certain that the button to the left of **Global** at the top right of the dialog is "on"; if not, *click* on it. Exit the dialog by *clicking* on **Save As** and supply a new name "cyclooctene.cis_am1". What you have asked for is an optimization at the AM1 level on each of the different conformers of *cis*-cyclooctene.

6. Submit the job (set of individual geometry optimizations). When completed (it will take a few minutes), examine the AM1 heats of formation of the individual conformers. You can sort the list from lowest to highest energy by selecting **by Energy** under the **Sort** menu in the **Lists** dialog.

7. Repeat steps 1 to 6 for *trans* cyclooctene. Name the MM3 job: "cyclooctene.trans_mm3" and the AM1 job: "cyclooctene.trans_am1". Calculate the *cis-trans* energy difference at both MM3 and AM1 levels (assuming the lowest-energy conformer for each). Do molecular mechanics and semi-empirical calculations yield similar results?

8 and 9 optional

8. Let's see if the ordering of conformer stabilities changes in going from the gas phase into a non-polar solvent (hexadecane). It is not necessary to repeat the (AM1) geometry optimizations, but only to perform single energy solvent calculations.

   Make certain that both "cyclooctene.cis_am1" and "cyclooctene.trans_am1" are on screen; if not read them using **Open** from the **File** menu. For each set of *cis* and *trans* cyclooctene conformers, enter the **Semi-Empirical** dialog, select **Single Point Energy** under **Task** and **Hexadecane** under **Solvent** (**AM1** still should be selected under **Model**). Again, make certain that the button to the left of **Global** is set. *Click* on **Save As** and provide names "cyclooctene.cis_am1hd" and "cyclooctene.trans_am1hd", respectively.

9. Submit the jobs. When completed, select the lowest for each of the *cis* and *trans* systems and compute the difference in energies (based on the lowest-energy conformers). Has solvent altered the *cis-trans* ordering seen in the gas phase?

10 to 12 optional

Different conformers are more easily compared if they are optimally aligned according to their structures. Let's do this for one of the data sets which you have previously generated.

10. Select the lowest-energy conformer for one of the sets of cyclooctene conformers which you have previously generated, e.g., "cyclooctene.cis_am1".

11. Select **Superposition/Similarity** under the **Build** menu.

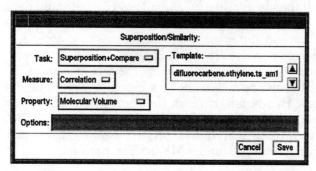

Select **Superposition+Compare** under **Task, Correlation** under **Measure** and **Molecular Volume** under **Property**. You have requested that the cyclooctene conformers all be aligned (with the lowest-energy conformer acting as a template) based on best fit of their van der Waals representations. *Click* on **Save** to exit the dialog.

12. Submit the job. Once it has completed, examine the different conformers a pair at a time. To do this, *click* on the button to the left of **Multiple** (**Lists** dialog). Two (or more) conformers can then be simultaneously displayed by selecting (*clicking* on) their respective entries from the list. *Clicking* again on the list entry deselects the conformer.

15. Close all active files (**Close** under the **File** menu).

Vinyl, Allyl Ether

Vinyl,allyl ether is the reactant in the parent Claisen rearrangement,

the concerted mechanism for which demands a *chair* arrangement of the reactant. Is this the lowest-energy conformer of the molecule, or must a significant energetic penalty be paid in order to properly orient the molecule for reaction? Let's perform a conformation search using the PM3 semiempirical model to quantify the energy of the best *chair* structure relative to the energy of the global minimum. The following steps are required:

1. Enter the builder. Construct vinyl,allyl ether (in any conformation) from a sequence of sp^2 and sp^3 carbons and an sp^3 oxygen. Minimize

and exit the builder (**Quit** from the **File** menu). Supply the name: "vinylallylether_pm3".

2. Select **Conformer Search** from the **Build** menu. Once inside the dialog, *double click* in turn on the three "internal" single bonds. Each bond will then be marked by a small gold cylinder. By default, the step size is 120°. This is a good choice for rotation about the bond connecting sp^3 carbon and sp^3 oxygen but, in order to save time, can be changed to a larger step size (180°) for rotation about the other two single bonds involving sp^2 carbon. To change the default value of "**3**" (meaning 360°/3), which appears in the text box to the right of **Fold Rotation** for these two bonds, *click* on the appropriate bond, position the cursor inside the text box, delete the "**3**" and type in "**2**". Before you exit the **Conformer Search** dialog (**Quit** from the **File** menu), make certain that **Systematic** (under **Method**) is specified.

3. Select **Semi-Empirical** from the **Setup** menu. Specify **Geometry Optimization** for **Task**, **PM3** for **Method** and **None** for **Solvent**. The default values of **Charge** (0) and **Multiplicity** (1) are correct. Exit the dialog by *clicking* on **Save**.

4. Submit the job. When completed, examine the individual conformers and their energies by *clicking* on the associated entries given in the **Lists** dialog (you need to select **Show Energy** under the **View** menu in the **Lists** dialog). See if you can identify one or more conformers which are in *chair* arrangements and hence suitable candidates for the Claisen rearrangement.

5 to 7 optional

You can obtain a better estimate of energy required for conformation reorganization by performing *ab initio* calculations.

5. Sort the list of conformers by energy (select **by Energy** from the **Sort** menu in the **Lists** dialog), and then delete from the list all but the lowest-energy conformer and the lowest-energy conformer which adopts a *chair* arrangement (**Delete** from the **Member** menu in the **Lists** dialog). When you are done the list should contain only two members.

6. Enter the **Ab Initio** dialog (**Setup** menu) and specify **Single Point Energy** for **Task**, **HF** for **Theory** and **3-21G$^{(*)}$** for **Basis**. The default values of **Charge** (0) and **Multiplicity** (1) are correct. Make certain that the **Direct** button is on. These calculations are large enough that they may require more memory than is available on your workstation. You can save time by using the wavefunctions from the previous PM3 calculations as guesses for the *ab initio* calculations. *Click* on **Wavefunction** to the right of **Restart Using** at the bottom of the dialog. *Click* on the button to the left of **Global** at the top right of the dialog. Exit the dialog by *clicking* on **Save As** and supplying the name: "vinylallylether_321g".

7. Submit the job. When completed (it will require several minutes) examine the 3-21G conformer energies (use **Show Energy** under the **View** menu in the **Lists** dialog). Estimate the penalty which must be "paid" in order to go from the ground-state conformer to one which is properly poised to undergo the Claisen rearrangement.

8. Close all active files (**Close** under the **File** menu).

2.5

Building and Locating Transition States

Several factors complicate obtaining transition states even for very simple chemical reactions. Foremost among them is an almost complete lack of knowledge of the detailed geometrical structures of transition states. Unlike "normal" molecules, where reasonable valence structures may generally be written and, based on these structures, accurate guesses at bond lengths and angles may be made, designation of appropriate valence structures for transition states, let alone detailed geometrical parameters, is not straightforward. There are, however, emerging strategies both for construction of transition states for "new" reactions, and for use of previous information obtained at low levels of calculation and/or on model reactions. The examples below illustrate some of these strategies. Additional discussion is provided in **Section 5.2** and elsewhere[5].

Pyrolysis of Ethyl Formate

Ester pyrolysis offers a convenient way to introduce a CC double bond into a molecule. The mechanism involves simultaneous transfer of a hydrogen atom to the carbonyl oxygen along with CO bond cleavage, e.g., in ethyl formate.

ethyl formate   ethylene   formic acid

Let's use the **Linear Synchronous Transit** (LST) method to produce a guess at the transition state for pyrolysis of ethyl formate, and then obtain the transition state at the AM1 semi-empirical level. We'll represent the product (ethylene + formic acid) as a weak complex, which can be built from the reactant by constraining a number of key bond distances. Given reactant and complex (product) structures, we'll use the linear synchronous transit method to provide a guess at the transition-state geometry, and then locate and verify the transition state.

1. Enter the builder. Build ethyl formate (start from **Ester** from the **Groups** menu). Arrange in a conformation in which one of the terminal hydrogens on the ethyl group is "poised" to transfer to the carbonyl oxygen. Minimize, select **Save As** from the **File** menu (inside the builder), and name the job: "ethylformate_am1".

2. Make a copy of the file using **Save As** under the **File** menu (you do not need to leave the builder). Name this: "formicacid.ethylene_am1". Select **Constrain Distance** under the **Geometry** menu. This leads to a message

> Constrain Distance: Select 2 atoms, a bond or a distance constraint. (Enter "." to abort)

One after the other, *click* on atoms "a" and "b" in the figure below.

This defines the CH bond which will break in the pyrolysis, and leads to a dialog box.

Click on the button to the left of **Constrain Distance,** and replace whatever value appears in the text box immediately to the right with **2.1** (much longer than a normal CH bond). *Click* on **OK**.

In a similar way, constrain the atoms "b" and "c" to be separated by a distance of 1.1 and the atoms "d" and "e" to be separated by a distance of 2.4. *Click* on **Minimize**; in a few seconds a structure with the three constraints you have imposed (approximately) enforced will be produced. The resulting structure will serve as the "product" in the pyrolysis reaction.

3. Exit the builder (**Quit** from the **File** menu). Bring the reactant "ethylformate_am1" back on screen (**Open** from the **File** menu).

4. *Click* on "ethylformate_am1" to make it the active molecule. It will become the "reactant". Select **Transition Search** from the **Build** menu. A message will appear in the menu bar at the top of the screen:

Transition Search: Select a molecule to specify the product. (Enter "." to abort)

Click anywhere on the structural model for "formicacid.ethylene_am1" to designate it as the "product". A new dialog will appear,

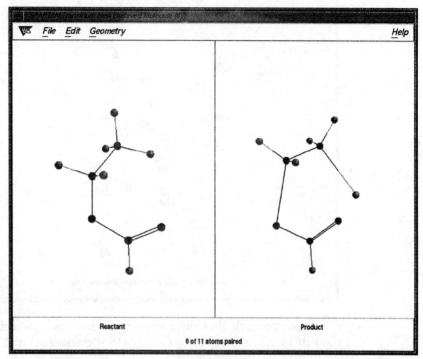

with ethyl formate (the reactant) on the left and the formic acid/ethylene complex (the product) on the right. You now need to associate "equivalent" atoms on the reactant and product. This can be made easier by reorienting one or both molecules such that the atoms on one molecule may be (visually) associated with equivalent atoms on the other molecule (as above).

Association occurs by first *clicking* on the ball designating an atom on one molecule, followed by *clicking* on the ball designating the equivalent atom on the second molecule to which the first atom is to be paired. Upon selection of the first atom of a pair, the associated ball will be colored gold; *clicking* on this atom a second time (without *clicking* on an atom in the second molecule) deselects it, returning the ball to its original color. Another atom (in either molecule) may then be selected. Upon selection of an associated atom, both balls will disappear. Attempts to *click* in succession on two atoms on the same molecule, or to *click* on an atom on the second molecule of different atomic number, will result in a bell, signaling that pairing is not permitted. Try again.

This procedure needs to be repeated until all atoms are paired. When this is completed, select **Generate** from the **Edit** menu. After a few seconds, the "split screen" containing the reactant and product molecules will be replaced by a screen containing the guessed transition state.

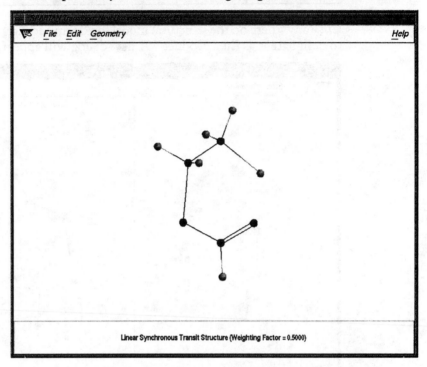

Linear Synchronous Transit Structure (Weighting Factor = 0.5000)

Convince yourself that the structure is reasonable, select **Quit** from the **File** menu, and supply a name: "ethylformate.ts_am1".

5. Enter the **Semi-Empirical** dialog and specify **Transition Structure** for **Task, AM1** for **Model** and **None** for **Solvent**. The default **Charge** (0) and **Multiplicity** (1) values are correct. Type "**optcycle = 300**" in the text box to the right of the word **Options;** this is probably not necessary, but in case your guess was poor, will allow an ample number of optimization cycles for the search to complete. *Click* on **Save** to exit the dialog. Before you submit the job, enter the **Properties** dialog (under the **Setup** menu) and *click* on the button to the left of the word **Frequency.** This requests a normal-mode analysis following the optimization. This is needed to confirm that you have found a transition state, and to allow animation along the reaction coordinate to ensure that this transition state smoothly connects reactant and product. *Click* on **Save** to exit the dialog.

6. Submit the job. It may require a few minutes to complete; transition state optimization typically requires several times the number of optimization steps needed for optimization of equilibrium geometry. In the meantime, you can remove from the screen the two molecules used to build the transition state guess, "ethylformate_am1" and "formicacid.ethylene_am1" (**Close** from the **File** menu). When the job has finished, examine both the geometry of the transition state (using functions under the **Geometry** menu), and then animate the motion

of atoms around the transition state. Select **Vibration** under the **Display** menu,

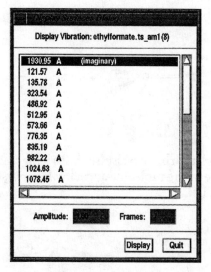

click on the top entry in the large text box (there should be indication that it corresponds to an imaginary frequency) and *click* on **Display**. Make certain that the motion is consistent with the reaction of interest and not with some other process. When you are finished, turn off the animation by reentering the **Vibration** dialog and *clicking* on **Quit**.

7 and 8 optional

7. A density isosurface permits us to examine the distribution of electrons in this simple transition state. Enter the **Surfaces** dialog (**Setup** menu), and specify **density (bond)** for **Surface**, **none** for **Property** and **med** for **Resolution**. *Click* on **Add** and then on **Save** to exit the dialog.

8. Submit the job. When completed, enter the **Surfaces** dialog (**Display** menu). Highlight (*click* on) the entry "surface = density, ...", *click* on the button to the left of **Display Surface** and select **Solid** under **Style**. *Click* on **OK** to exit the dialog.

    You should be able to convince yourself that the CO bond in ethyl formate has nearly fully cleaved and that the migrating hydrogen is midway between carbon and oxygen. If you like, replace the opaque solid representation of the density surface by a transparent solid or by a mesh in order to see the underlying skeleton. You will need to reenter the **Surfaces** dialog (**Display** menu).

9. Close all files (**Close** from the **File** menu).

Addition of Singlet Difluorocarbene to Ethylene and to Cyclohexene

Singlet carbenes add to olefins to yield cyclopropanes.

Since a singlet carbene possesses both a high-energy filled molecular orbital in the plane of the molecule and a low-energy, out-of-plane unfilled molecular orbital, i.e.,

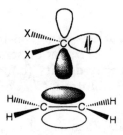

this reaction presents an interesting dilemma. Clearly it would be more advantageous for the low-lying vacant orbital on the carbene, and not the high-lying filled orbital, to interact with the olefin π system during its approach, i.e.,

although, of course, this leads to a product with an incorrect geometry. It must be that the carbene "twists" by 90° during the course of reaction, i.e.,

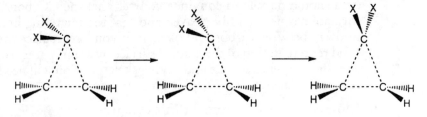

In this example, we'll use the distorted cyclopropane,

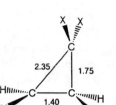

as a guess for the transition state, which we'll then optimize using the semi-empirical AM1 model. This will allow us to see if our qualitative thinking regarding the approach of the carbene stands up to the quantitative molecular orbital calculations. Perform the following steps:

1. Enter the builder. Place cyclopropane (from the **Rings** menu) on screen, and add two fluorines to one of the carbons.

2. Select **Constrain Distance** from the **Geometry** menu. A message will appear at the top of the screen instructing you to identify the constraint. *Click* on one of the CC bonds involving the CF_2 group. *Click* on the box to the left of **Constrain Distance**. A text box will appear indicating the current bond length (1.5588Å) and providing the opportunity to introduce a constraint. *Click* inside the text box to the right of **Constrain Distance**. Change the contents of the text box from **1.5588** to **1.75**. *Click* on **OK**. Repeat the procedure to constrain the two remaining CC bond distances as indicated in the figure above.

3. *Click* on **Minimize** to produce a refined structure subject to these constraints. Exit the builder and supply the name: "difluorocarbene.ethylene.ts_am1".

4. Specify **Transition Structure** under **Task**, **AM1** under **Model** and **None** under **Solvent** in the **Semi-Empirical** dialog. The default settings for **Charge** (0) and **Multiplicity** (1) are correct. In the text box to the right of **Options**, type "optcycle=200 nosymtry". It's almost always a good idea to turn off symmetry during transition state optimizations; we really do not know what transition states look like and it is better not to bias the search.

   Exit the **Semi-Empirical** dialog by *clicking* on **Save**, and enter the **Properties** dialog (**Setup** menu). *Click* inside the box to the left of **Frequency**. This will allow you to verify that you have located a transition structure, and furthermore, allow animation of the motion along the reaction coordinate. Exit the dialog by *clicking* on **Save**, and submit the job.

5. When the job is complete, examine the geometry of the transition structure. In light of considerations regarding the orientation of filled and empty molecular orbitals on the carbene and the π orbital on ethylene, would you describe your structure as corresponding to an "early" or "late" transition state?

6. Animate the vibrational mode corresponding to the reaction coordinate. Select **Vibration** under the **Display** menu. *Click* on the imaginary fre-

quency at the top of the list of frequencies (this corresponds to motion along the reaction coordinate), then *click* on **Display.** Does the animation show reorientation of the carbene as it approaches the olefin in line with the qualitative picture? When you are finished, turn off the animation by reentering the **Vibration** dialog and *clicking* on **Quit.**

7 optional

7. Select **Isotopes** under the **Build** menu. A ball-and-stick model will appear with "**1**" beside each hydrogen, and "**12**" beside each carbon. These are the default hydrogen and carbon masses, respectively. In turn, *click* on each of the hydrogens; the numbers (masses) will change to "**2**". (*Clicking* again will change the mass to "**3**" and once more back to "**1**".) When completed, *click* on **Save** in the dialog box at the top left of the screen. Resubmit the job, and when complete examine the new set of vibrational frequencies either from the text output, or from the vibration dialog.

8 to 12 optional

8. Select **Vibration Sequence** under the **Build** menu. The dialog which results,

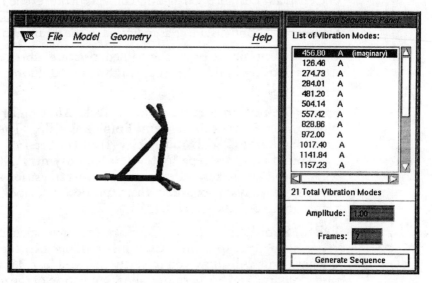

is divided into two regions: a *work area* on the left for molecule display, and a *control area* on the right. *Click* on the imaginary frequency inside the text box at the top of the *control area*, and then *click* on the **Generate Sequence** button. This will generate a continuous sequence of structures along the reaction coordinate with the transition state in the middle. (The amplitude of motion and the number of frames in the sequence can be adjusted if desired by changing the default values inside the text boxes to the right of **Amplitude** and **Frames**, respectively, and again *clicking* on **Generate Sequence**.) Exit the dialog by selecting **Quit** from the **File** menu, and supply the name: "difluorocarbene.ethylene.ts.sequence_am1".

9. You have made a list, and associated with it is the **Lists** dialog, demarking the individual "frames". Enter the **Surfaces** dialog (**Setup** menu) and specify calculation of two surfaces: **density (bond)** and **HOMO**. *Click* on the button to the left of **Global** at the top right hand corner of the dialog. *Click* on **Save** to exit the dialog. What you have done is to request calculation of these two surfaces for the entire set of frames associated with the vibration.

10. Submit the job. When completed, enter the **Surfaces** dialog (**Display** menu), *click* on the text string "surface = density, ...", *click* on the button to the left of **Display Surface**, select **Solid** under **Style** and finally *click* on **OK** to exit the dialog. A single frame will appear. Animate the display by *clicking* on **Animate** (**Lists** dialog), or step through the individual frames using the keys to the left of **Step** (**Lists** dialog).

    Experiment with various graphical displays (mesh and transparent solid will allow you to see the underlying skeleton), and with different structural models. Perhaps the most "truthful" representation is a solid surface and no model at all (**Hide** under the **Model** menu). Note from the animation that the density moves smoothly from two separated charge distributions (representing difluorocarbene and ethylene) to a single charge distribution (representing 1,1-difluorocyclopropane).

11. Reenter the **Surfaces** dialog (**Display** menu), turn off display of the total density (*click* on the entry "surface = density..." and *click* on the button to the left of **Display Surface**) and turn on display of the HOMO (*click* on the entry "surface = homo...", *click* on the button to the left of **Display Surface** and select **Solid** under **Style**). *Click* on **OK** to exit the dialog and once in the main window, step through the frames one-by-one. Focus on one part of the display, e.g., one lobe of the π bond in ethylene. You may see the color of this lobe change (from red to blue, or vice versa) as you step through the frames. This is unavoidable, and is a consequence of the fact that the sign of the molecular orbital coefficients (within a given molecular orbital) is arbitrary. "Swap" the phase of particular frames (**Swap MO Phase** button) so that the progression is uniform. When you are done, animate the sequence of frames; notice the progression of images from one extreme, displaying the lone pair on the carbene and the π orbital on ethylene, to the other extreme, displaying one of the valence (Walsh) orbitals of the cyclopropane.

12. Remove the image from the screen (**Close** under the **File** menu).

13 to 17 optional

13. If it is not already on screen, bring up the original file "difluorocarbene.ethylene.ts_am1" (**Open** under the **File** menu). Make a copy (**Save As** under the **File** menu), and name it: "difluorocarbene.cyclohexene.ts._am1".

14. Enter the builder (**Edit Structure** under the **Build** menu). Add four sp^3 carbons to one side of the "ethylene" fragment and bond together to make a "cyclohexene" ring. **Do not minimize.** Select **Freeze Cen-**

ter from the **Geometry** menu. *Click* on **Freeze All** in the text box which appears. Magenta colored markers will appear on all atoms (and free valences); these indicate atoms which will not move in the optimization. *Click* (only) on the four sp³ carbons which you have just added (and on the two free valences associated with each) to remove the markers from these positions. What you have done is to specify that the underlying transition structure is not to be altered during optimization (in the builder), but the modifications which you have made (changing from "ethylene" to "cyclohexene") are to be refined. *Click* on **Done** to exit the **Freeze Atom** dialog, and then **Minimize** to obtain a refined structure. Exit the builder by selecting **Quit** from the **File** menu.

15. The job is already set up for a transition structure optimization at the semi-empirical AM1 level and for a normal-mode analysis. All you need to do is to submit the job.

16. When completed, examine the text output (**Output** under the **Display** menu). You should see that the total number of cycles required for the transition structure optimization is relatively small, in view of the complexity of the system. This reflects the very good guess at the structure. You can animate the motion corresponding to the reaction coordinate to make certain that it indeed corresponds to difluorocarbene addition.

17. Close all files (**Close** under the **File** menu).

2.6

Building Polypeptides

The problem of single-bond conformation is never more apparent than in the geometries of polypeptides. Here, distinct local domains involving helices and sheets (among other structures) occur commonly, and these in turn dictate overall macroscopic geometry. For sufficiently large polypeptides, i.e., proteins, only one or a very few conformers (out of an astronomical number of possibilities) dominate, leading in effect to highly rigid geometries. This is certainly a major factor behind the ability of proteins to direct specific chemical reactions.

Glycine Tripeptide

We'll use the most simple tripeptide both as a non-zwitterion and as a zwitterion to illustrate the basics of polypeptide construction inside of SPARTAN. Following optimization at the SYBYL molecular mechanics level, we'll perform AM1 and $AM1_{aq}$ semi-empirical calculations to assess the role of the solvent in altering tautomer stabilities.

1. Get into the *peptide* builder (by going into the *entry* builder, and *clicking* on **Peptide**).

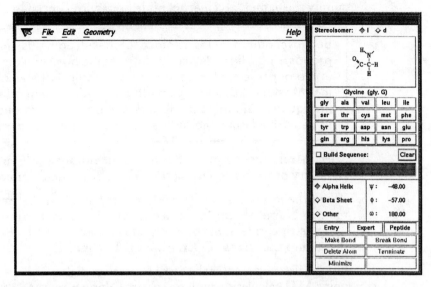

This dialog, like the other builder screens, is made up of two areas, a *model kit* on the right and a *work area* on the left.

2. *Click* on the button to the left of **Build Sequence,** and then *click* three times on "gly" (glycine) from the list of amino acid template codes. The structure of glycine will appear in the large text box at the top of the *model kit,* and the sequence in the smaller text box underneath the **Build Sequence** button.

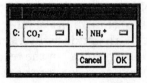

3. *Click* on the button to the left of **Alpha Helix.** The backbone ψ and ϕ angles which define an a helix will appear at the right. Move the cursor into the *work area* and *click* once. The skeleton of triglycine will appear, although it will not be properly terminated.

4. Let us first work with the non-zwitterionic structure. *Click* on the **Terminate** button in the *model kit.* This will result in display of a small dialog in the *work area.*

Select CO_2H from the menu to the right of **C** and NH_2 from the menu to the right of **N,** and then *click* on **OK** to close the dialog and terminate the polypeptide. Exit the builder (**Quit** from the **File** menu), and supply the name: "glycinetripeptide.nonzwitterion_sybyl".

> Because amino acids replace atomic fragments, functional groups, rings, ligands and chelates as the basic building blocks in the *peptide* builder, these other building blocks are missing. Most modifications of peptides, aside from modifications in sequence of amino acids and in overall conformation, need to be carried out either in the *entry* or *expert* builder screens.

5. Select **Mechanics** from the **Setup** menu. Supply a title, and specify a geometry optimization using the SYBYL force field. Submit the job.

6. When the job is complete, make two copies (**Save As** from the **File** menu). Name them: "gylcinetripeptide.nonzwitterion.am1" and "gylcinetripeptide.nonzwitterion_ am1aq", respectively. Bring both of them onto the screen (**Open** from the **File** menu).

7. For the first copy, enter the **Semi-Empirical** dialog, and specify a single point AM1 calculation with no solvent (**None** from the **Solvent** menu). For the second copy, specify a single point calculation with the $AM1_{aq}$ model (select **Water** from the **Solvent** menu). The default **Charge** (0) and **Multiplicity** (1) are correct for both calculations. Submit both calculations.

8. As each job completes, select **Properties** under the **Display** menu and then **Energy** from the sub-menu which appears. Write down the calculated heat of formation (in kcal/mol) in the table below.

	non-zwitterion	zwitterion	Δ
gas ΔH_1(AM1)			
water ΔH_1($AM1_{aq}$)			

9. Repeat all your calculations, this time using the zwitterionic structure for glycine tripeptide. This is accomplished by building glycine tripeptide as you did before, but this time selecting NH_3^+ from the available "N-terminator" groups, and CO_2^- from the "C-terminator" groups. Name the mechanics job: "glycinetripeptide.zwitterion_sybyl", and the two semi-empirical jobs: "glycinetripeptide.zwitterion_am1" and "glycinetripeptide.zwitterion_am1aq", respectively. You need to specify **"ionicgroups"** in the text box to the right of **Options** in the **Semi-Empirical** dialog for the $AM1_{aq}$ calculation on the zwitterion; this instructs SPARTAN to look for "ionic" functionality (O^- and NH_3^+ groups), even though the molecule is neutral. When the semi-empirical calculations are both done, write down the heats of formation in the table above, and calculate the difference in stabilities between neutral and zwitterionic forms. According to this model, which form of triglycerine (neutral or zwitterionic) is the more stable in the gas phase? Which is the more stable in water?

10. Close all molecules on the screen.

11 to 15 optional

Let's illustrate that, according to MM3 molecular mechanics, the non-zwitterionic form of glycine tripeptide does not prefer a helix in the gas phase, but rather uncoils. We'll also use this example to illustrate the use of genetic algorithms for conformation searching.

11. Build glycine tripeptide as an a helix (as you did previously). Terminate it with NH_2 and CO_2H groups and name it: "glycinetripeptide.nonzwitterion_mm3".

12. Enter the **Conformer Search** dialog (**Build** menu). One after the other *double click* on the seven single bonds indicated by bold lines in the schematic below.

$$H_2N-CH_2-\overset{\overset{O}{\|}}{C}-NH-CH_2-\overset{\overset{O}{\|}}{C}-NH-CH_2-\underset{\underset{O}{\diagdown}}{\overset{O-H}{C}}$$

This includes all single bonds except the two amide linkages which are assumed to be rigid. Each of the bonds should be marked by a gold cylinder. Select **Genetic Algorithm** from the available **Methods**, and then exit the dialog by selecting **Quit** from the **File** menu.

> Of the conformer searching methods available in SPARTAN, only the Monte Carlo and genetic algorithm methods may be applied to systems with large numbers of flexible rotors.

14. Enter the **Mechanics** dialog (**Setup** menu). To save time we won't optimize the geometry at every point in the conformation search but rather assume "rigid rotation". Type in a title, specify **Single Point Energy** for **Task** and **MM3** for **Force Field**. Exit the dialog by *clicking* on **Save**.

15. Submit the job. When it has completed, examine the individual conformers by selecting them from the **Lists** dialog. Is the lowest energy form still an a helix?

16. Close "glycinetripeptide.nonzwitterion_mm3".

3

Electronic Structure Models

Here we provide a brief and non-mathematical description of *ab initio* (Hartree-Fock and correlated), local density functional and semi-empirical electronic structure models. Our primary purpose is to provide the reader with some level of insight into the origins of the various methods and the relationships among them. More thorough treatments have been presented elsewhere.[3,7,8]

3.1

Ab Initio Models

Ab initio models may be subdivided into two categories: Hartree-Fock models and correlated models. Three approximations take us from the full many-electron Schrödinger equation to practical Hartree-Fock models:

1. Separation of nuclear and electron motions (the **Born Oppenheimer approximation**). In effect, what this says is that "from the point of view of the electrons the nuclei are stationary". This eliminates the nuclear kinetic energy term in the Hamiltonian and leads to a constant nuclear-nuclear potential energy term. In so doing, it eliminates the mass dependence in what is now referred to as the electronic Schrödinger equation.

2. Separation of electron motions (the **Hartree-Fock approximation**). What is actually done is to represent the many-electron wavefunction as a sum of products (in the form of a determinant) of one-electron wavefunctions, termed molecular orbitals.

3. Representation of the individual molecular orbitals in terms of linear combinations of atom-centered basis functions or atomic orbitals (the **LCAO approximation**). This reduces the problem of finding the best functional form to a much simpler problem of finding the best set of linear coefficients.

These three approximations lead to the Roothaan-Hall equations, the computation cost of which scales as the fourth power of the total number of atomic orbitals. These equations are not solvable in closed form; rather their solution requires an iterative self-consistent-field (SCF) procedure.

Most practical correlated methods start from the best Hartree-Fock wavefunction (for a given basis set) and then proceed to "relax" the second approximation (separation of electron motions). Operationally, this is accomplished either by implicit or explicit mixing of the ground-state (lowest-energy) determinant with excited-state (higher-energy) determinants. In the limit of full mixing with a complete basis set (something that is not practical), this is equivalent to full coupling of electron motions.

Ab initio methods may be thought of in terms of a two-dimensional chart (**Figure 3-1**), the horizontal dimension corresponding to the extent to which electron motions are coupled, and the vertical dimension corresponding to the size of the basis set. Hartree-Fock models constitute the leftmost column on the chart, that is, full separation of electron motions. Here, the limit of a complete basis set is termed the Hartree-Fock limit, which is not the same as the exact solution of the full many-electron Schrödinger equation (or experiment).

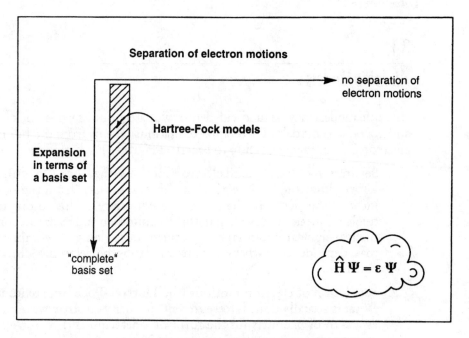

Figure 3-1. Designation of *ab initio* methods

In practice, it is not possible to reach the lower right-hand corner of this diagram, that is, to solve the many-electron Schrödinger equation, at least for real systems. We can, however, think of a situation where we successively improve our treatment by systematic progression in the two dimensions of the chart, that is, by more and more relaxing the two approximations. In the event that we reach a point where further improvements do not lead to further changes in the property of interest for the system of interest, then we can imply that we have actually "solved" the electronic Schrödinger equation in this limited domain. That is to say, we can use convergence of a particular property to judge its accuracy without recourse to experiment. Thus, we can deal with molecules which may not be sufficiently stable to be investigated experimentally or which may not even exist, e.g., transition states, as well as analyze quantities which are not experimental observables, e.g., atomic charges.

What remains to be specified for both *ab initio* Hartree-Fock and correlated calculations is the nature of the atomic basis set, both in terms of the kind and number of functions. Here, we provide just a few details about a number of basis sets in widespread use.

Essentially all present day molecular orbital methods make use of Gaussian basis functions, written in terms of polynomials in x, y, z multiplying an exponential in r^2, i.e.,

$$x^l y^m z^n \exp(\alpha r^2),$$

where α is a constant which controls the radial extent of the function. These functions are by convention named s, p, d, etc., depending on the order of the polynomial. The sum of the integers l, m, n is 0 for an s function, 1 for a p function, 2 for a d function, etc.

The simplest (smallest) possible atomic orbital representation is termed a **minimal basis set**. This comprises only that number of functions required to accommodate all of the electrons of the atom, while still maintaining overall spherical symmetry. This requires a single (1s) function for hydrogen and helium, two functions (1s, 2s) for lithium and beryllium, and a set of five functions (1s, 2s, $2p_x$, $2p_y$, $2p_z$) for the remaining first-row elements (boron to neon). Similarly, the minimal basis description of sodium and magnesium is made up of six functions (1s, 2s, $2p_x$, $2p_y$, $2p_z$, 3s), and of the rest of the second-row elements of nine functions (1s, 2s, $2p_x$, $2p_y$, $2p_z$, 3s, $3p_x$, $3p_y$, $3p_z$). Practical minimal basis sets such as the popular STO-3G representation, however, supplement their descriptions of lithium and beryllium (and heavier alkali and alkaline-earth elements) by a set of unoccupied (in the atom) but energetically low-lying p-type functions, e.g., 2p in the case of lithium and beryllium. These have been found to be necessary to accommodate back-bonding in molecular systems.

Minimal basis sets are not capable of adequately describing non-spherical (anisotropic) electron distributions in molecules. This is because the individual basis functions are either themselves spherical (s functions), or come in sets of equivalent functions which uniformly span a spherical space (p and d functions). The simplest remedy for any problems which such a restriction causes is to "split" the valence description into "inner" and "outer" components. In a so-called **split-valence basis set**, the valence manifolds of main-group elements are represented by two complete sets of s- and p-type functions. You can think of the effect this has by recognizing that the coefficients which multiply the individual basis functions are "variables" in the calculation. For example, to describe a p function which is to be used in construction of a "tight" σ bond, one needs to mix a large amount of a highly contracted basis function with a small amount of a more diffuse function,

$$p_\sigma = \text{large} * \quad \text{+ small} * \quad \Rightarrow \quad$$

while the reverse would be true for construction of a p function to be employed in making a "looser" π bond.

$$p_\pi = \text{small} * \quad \text{+ large} * \quad \Rightarrow$$

For completeness, practical split-valence basis sets represent hydrogen in terms of a pair of (contracted and diffuse) s-type basis functions.

Among the most simple split-valence basis sets is 3-21G. While this performs quite well for structure and energetic comparisons involving molecules comprising hydrogen and first-row elements only, it generally performs poorly in the description of molecules incorporating second-row and heavier main-group elements. Adequate descriptions here have been found to require incorporation of unoccupied (in the atom) but energetically low-lying d-type functions, a situation reminiscent of descriptions of alkali and alkaline-earth elements which require unoccupied p-type functions. A very simple representation of this type is the 3-21G$^{(*)}$ basis set, which is used in conjunction with 3-21G for hydrogen and first-row elements.

d-functions on main-group elements (where they are not occupied) also allow for small displacements of the center of electronic charge contained in valence p-type functions away from the nuclear positions (on which the basis functions are centered).

$$\quad + \lambda \quad \Rightarrow$$

This is very closely related to Pauling's hybridization arguments, where the common example relates to the construction of an "sp hybrid" from mixture of a p-type function into a valence s orbital.

$$\quad + \lambda \quad \Rightarrow$$

Functions added to basis sets in order to achieve displacement are generally termed polarization functions, and basis sets which incorporate polarization functions are called polarization basis sets. Among the popular polarization basis sets in widespread use are the 6-31G* and 6-31G** representations, which include polarization functions only on heavy atoms and on all atoms, respectively.

Larger basis sets involving multiple sets of valence functions and/or multiple sets of polarization functions, and possibly as well functions of even higher angular quantum number, or functions which extend very far from the nuclear positions (diffuse functions), have also been employed, but rapidly become very expensive for routine applications.

3.2

Density Functional Models

One way to think about density functional theory is to consider the description of the total energy of a molecular system from the standpoint of Hartree-Fock theory. This involves the kinetic energy of the individual electrons (recall that the nuclear kinetic energy is zero within the Born-Oppenheimer approximation), the repulsive potential energy between nuclei (a constant for a given nuclear configuration), the attractive potential energy between nuclei and electrons, and finally the repulsive potential energy between electrons.

Two terms involving electron-electron interaction actually appear in the Hartree-Fock Hamiltonian. One of these clearly contributes in a repulsive manner, and is labeled the Coulomb term. The other term appears to contribute in an attractive manner, and therefore cannot easily be given a classical physical interpretation, except to say that it somehow compensates for the apparent overestimation of electron-electron repulsion given by the Coulomb term. This second term is called the exchange term.

Note that the (limiting) Hartree-Fock energy is not the experimental energy, that is, it does not correspond to the energy which follows from exact solution of the many-electron Schrödinger equation. This is because of the "separation of electron motions approximation" (see **Section 3.1**), which does not allow the individual electrons sufficient flexibility "to get out of each other's way". The difference between the Hartree-Fock energy and the experimental energy is termed the correlation energy, meaning the energy resulting upon correlation of the motions of the electrons.

Density functional theory takes the electron kinetic term, the nuclear-nuclear and nuclear-electron potential terms and the Coulomb term directly from Hartree-Fock theory. It replaces the exchange term and adds a correlation term. (In the event that the exchange term is set to the Hartree-Fock exchange and the correlation term is set to zero, the density functional energy is the Hartree-Fock energy. In other words, Hartree-Fock theory may be thought of as a special case of density functional theory).

The exchange and correlation terms used in so-called local density functional models originate from the "exact" (numerical) solution of an idealized many-electron problem, mainly an electron gas of constant total electron density. While this may be a "reasonable" model for highly-extended systems, e.g., metals, its validity for calculations on molecules remains to be established. In practice, what is first done is to establish functional relationships between the exchange and correlation energies for such an "idealized gas" and the total electron density, i.e., to establish exchange and correlation functionals, and then to incorporate these relationships into the Hamiltonian describing a general many-electron system. Unfortunately, such relationships are not unique, and thus far no recipes guaranteeing "improvement" with systematic progression from one functional to another have been advanced.

Density functional models which allow for non-uniformity of the overall electron distribution (so-called non-local methods) have also been proposed. Also considered, have been "hybrid schemes", whereby the total exchange energy is arrived at by combining the Hartree-Fock exchange energy and the exchange energy resulting from some functional. One or more parameters adjusted to provide "best fit" with specific experimental observables may be utilized. Non-local density functional models and hybrid Hartree-Fock density functional models, like local density functional schemes, as yet offer no conspicuous paths for improvement.

Because the form of the correlation and exchange functionals are typically quite complex, they cannot be dealt with analytically, and practical density-functional calculations involve numerical integration steps. While these can lead to significant loss of precision (compared with *ab initio* methods which are strictly analytical), in principle the computational cost of a density functional calculation may be made to scale as the square of the number of basis functions times the number of integration points. The latter is a large number, and density functional calculations on very small systems will be slower than Hartree-Fock calculations, although the two methods will eventually cross in performance. In practice this crossover point is around 50-60 basis functions for the *ab initio* and density functional modules incorporated into SPARTAN.

Like Hartree-Fock and correlated *ab initio* calculations (and semi-empirical calculations as well), density functional calculations make use of explicit atomic basis sets. The same considerations apply in basis set selection as for these other methods, although in practice basis sets for density functional calculations are typically generated "on-the-fly" as numerical representations of best atom solutions.

3.3

Semi-Empirical Models

Semi-empirical models follow in a straightforward way from Hartree-Fock models. In effect, a single additional approximation, termed the NDDO approximation, is made. This is very severe, in that it eliminates overlap of atomic basis functions on different atoms (atomic basis functions on the same atom already do not overlap because of orthogonality). However, it leads directly to a reduction in computation effort from the fourth power of the total number of basis functions (in Hartree-Fock models) to the square of the number of basis functions. Other steps in practical semi-empirical models, e.g., matrix diagonalization, scale as the cube of the total number of basis functions and usually dominate the calculation.

Most present generation semi-empirical models are restricted to a minimal valence basis set of atomic functions. Inner-shell functions are not included explicitly, and because of this, the cost of doing a calculation involving a heavy main-group element, e.g., gallium, is no more than that incurred for the corresponding first-row element, e.g., boron. So-called Slater-type basis functions (STO's, which are closely related to the exact solutions for the hydrogen atom) are used in place of Gaussian functions employed for *ab initio* calculations.

Additional numerical approximations are invoked to simplify integral calculations, and more importantly adjustable parameters are introduced in order to reproduce specific experimental data as closely as possible. Choice of parameters is the key to successful semi-empirical methods.

Two of the three semi-empirical methods currently in widespread use (AM1 and PM3) are identical except for choice of parameters. AM1 was parameterized using a rather limited set of data thereby "guaranteeing" good results for these data, while parameterization of PM3 used a much larger and more diverse "training set".

The third and oldest method (MNDO) uses slightly different functional forms and fewer parameters. Among the newer semi-empirical methods is Thiel's MNDO/d, which supplements the valence descriptions of second-row and heavier main-group elements (only) by d-type functions (much like the relationship between 3-21G$^{(*)}$ basis sets for second-row and heavier main-group elements and 3-21G for first-row elements). Also recent is PM3 (tm), which describes transition metals in terms of a minimal valence basis set [nd, (n+1)s, (n+1)p, where n is the principal quantum number]. This is to be used in conjunction with PM3 for non-transition metals.

3.4

Selection of Appropriate Electronic Structure Models

Computational methods, in particular, electronic structure methods based on quantum mechanics, are now rapidly moving into the mainstream of chemistry. They are still very new, however, and the collective experience in their application to "real chemical problems" is still very limited. We need to ask some very basic questions about their reliability in preface to applications to new chemistry, and we need to be concerned with practical issues of whether or not the calculations can be performed sufficiently rapidly to actually be of use. All of this falls under a general heading of model selection.

Most important among many considerations involved in the selection of an appropriate electronic structure model are the level of confidence required in the results and the computational resources available. Only rarely will it be possible to utilize the most sophisticated theoretical treatment available, and even then this may not be sufficient to guarantee the level of accuracy desired. Typically, practical concerns will dictate use of lower levels of calculation. Thus, it is important for the user to understand in some depth the capabilities and limitations of available theoretical models, from semi-empirical models which can be applied to molecules comprising more than a hundred atoms, to the simplest minimal basis set Hartree-Fock treatments, which may be applied to systems comprising up to one hundred atoms, to correlated models (including density functional models) with large basis sets which are currently practical only for very simple molecules. Indeed, there will be problems where not even the simplest quantum chemical model can be used and only mechanics-based procedures are practical.

The assessment chapter which follows immediately provides a brief overview of the performance of the various electronic structure models, and should be of some use in defining the simplest level of calculation suitable for a particular problem at hand. Also useful for the purpose of model selection are "strategies" developed to attack specific classes of problems, typically by capitalizing on the strengths of particular models while carefully avoiding the weaknesses. Some discussion is provided in **Chapter 5**.

In the final analysis, there can be no substitute for experience, and this can be gained only through repeated applications.

4

Performance of Electronic Structure Models

Here we remark on the performance of *ab initio* and semi-empirical molecular orbital models and local density functional models with regard to the calculation of equilibrium and transition state geometries and reaction thermochemistry and kinetics. It is not our intention either to provide a thorough or critical assessment of the computational methods, but only a brief overview. More thorough discussion is available elsewhere.[3,4]

4.1

Equilibrium Geometries

Sufficient experimental structural data exist to allow thorough assessment of theoretical methods for the calculation of equilibrium geometries. While the majority of these pertain to molecules made up of light main-group elements (organic molecules), sufficient data exist for molecules incorporating heavier main-group elements and transition metals to allow some commentary.

Ab Initio Models

The following general conclusions may be drawn from the extensive comparisons which have been made:

1. Accurate equilibrium structures may be obtained from *ab initio* molecular orbital theory. Use of moderately large basis sets (6-31G* or larger) and Møller-Plesset treatment of electron correlation truncated at second-order (MP2) generally guarantees that calculated structural parameters will be close to measured equilibrium values. Smaller basis set MP2 calculations generally lead to bond distances which are longer than experimental values, sometimes significantly so, and cannot be recommended. Limiting MP2 bond lengths are typically, but not always, longer than experimental distances, although errors are quite small.

2. Hartree-Fock models using the 6-31G* polarization basis set generally overestimate bond lengths, typically by 0.01-0.02Å, although errors involving bonds between two highly electronegative elements can be much larger, e.g., ~ 0.1Å in F_2. These discrepancies reflect the behavior of the Hartree-Fock model rather than limitations in the underlying basis set. Limiting Hartree-Fock bond lengths are nearly always shorter than experimental values. Skeletal bond angles and dihedral angles calculated at the Hartree-Fock level are generally in good accord with experimental data.

3. Hartree-Fock models using small to medium size basis sets are generally successful in accounting for equilibrium structures of organic and (main-group) inorganic molecules. The 3-21G split-valence basis set (3-21G$^{(*)}$ for molecules incorporating second-row and heavier main-group elements) appears to be the simplest method of choice for widespread application. Even the STO-3G minimal basis set generally yields equilibrium geometries in good accord with experimental data.

4. Hartree-Fock models with minimal and split-valence basis sets do not provide a reliable account of the geometries of transition metal inorganics and organometallics, and cannot be recommended. There is insufficient experience with larger basis set Hartree-Fock models for systems incorporating transition metals to trace the origin of the deficiencies to limitations in the basis set or to insufficient treatment of correlation effects (or both).

Local Density Functional Models

There is much less experience with local density functional models, although some conclusions may be drawn regarding their performance:

1. Accurate equilibrium structures may be obtained from local density functional models with moderate to large basis sets including one or more sets of polarization functions. In the limit, local density functional models generally give bond lengths which are shorter than experimental values (typically by 0.01-0.02Å), the same result as noted for limiting Hartree-Fock models, but not the same as noted for comparable MP2 models.

2. Local density functional models with smaller (minimal and split-valence) basis sets without polarization functions almost always lead to bond distances which are longer than experimental values, often by as much as 0.1Å. They do not provide a reliable account of equilibrium geometry, and cannot be recommended.

3. Local density functional models provide a reasonable account of bond lengths in compounds incorporating heavy elements, including transition metal inorganics and organometallics. Large basis sets, with one or more sets of polarization functions and possibly including diffuse functions and f-type functions, are required for reliable descriptions.

Semi-Empirical Models

The following general conclusions may be drawn from the extensive comparisons both to experiment and to high-level *ab initio* calculations:

1. All three present-generation semi-empirical models (MNDO, AM1 and PM3) are generally suitable for the calculation of equilibrium geometries. All yield similar errors both for bond distances involving

heavy atoms and for skeletal bond angles. None of the models is as reliable as even the simplest *ab initio* schemes, and bond length and angle errors are typically twice as large as those resulting from Hartree-Fock 3-21G calculations.

2. The PM3 model is generally the best of the three schemes for structure determination. In particular, it is significantly better than MNDO and AM1 for the calculation of the geometries of molecules incorporating second-row and heavier main-group elements, most conspicuously for hypervalent molecules, i.e., molecules which exceed the normal complement of eight valence electrons. Limited data for MNDO/d suggests that it too is successful for the geometries of hypervalent molecules.

3. All semi-empirical models perform somewhat (but not markedly) less well for ions and free radicals than they do for "conventional" molecules. This is most likely due to the sparse representation of these types of molecules in the training sets used for parameterization, rather than to any inherent limitations of the models. MNDO and to a lesser extent AM1 models are unreliable in their description of hydrogen-bonded systems, while PM3 performs relatively well in this task.

4. Semi-empirical parameterizations for transition metals, based on a minimal valence basis set $[nd,(n+1)s,(n+1)p]$ and use of the underlying PM3 model for non-transition metals, lead to errors in bond lengths and bond angles involving metal centers on the order of 0.03-0.05Å and 3-5°, respectively, a similar range to errors for PM3 parameterizations for other elements.

5. Each of the methods has its own "quirks". For example, the PM3 model is known to improperly account for the planarity of amides and to show weak "attractions" between non-bonded hydrogens. In general, known problems are restricted to specific classes of molecules, and more than anything else, are probably a result of a lack of representation of these compounds in the original parameterizations.

Comparisons of Errors in Equilibrium Geometries among *Ab Initio*, Local Density Functional and Semi-Empirical Models

RMS errors in bond lengths and skeletal bond angles for molecules comprising first- or second-row elements, and which are well described in terms of conventional valence structures are provided in **Tables 4-1** and **4-2**, respectively. Other data sets would yield similar results, except that semi-empirical models may present difficulties for classes of molecules for which they have not been explicitly parameterized, e.g., charged species and radicals. Note, however, that situations where calculated structures at semi-empirical levels are outlandish are actually quite rare.

Table 4-1:   RMS Errors in Bond Distances Connecting Heavy Atoms (Ångstroms)

	STO-3G[a]	3-21G[(*)]	6-31G*
HF	0.042	0.028	0.028
MP2	—	0.044	0.048
SVWN[b]	—	0.027	0.018
MNDO	0.048	—	—
AM1	0.048	—	—
PM3	0.037	—	—

a) data for semi-empirical MNDO, AM1 and PM3 models refer to a minimal valence basis set of Slater-Type Orbitals (STO's).

b) local density functional.

Table 4-2:   RMS Errors in Skeletal Bond Angles (degrees)

	STO-3G[a]	3-21G[(*)]	6-31G*
HF	1.7	1.7	1.4
MP2	—	2.2	1.5
SVWN[b]	—	2.0	1.6
MNDO	4.3	—	—
AM1	3.3	—	—
PM3	3.9	—	—

a) data for semi-empirical MNDO, AM1 and PM3 models refer to a minimal valence basis set of Slater-Type Orbitals (STO's).

b) local density functional.

4.2

Transition State Geometries

The geometries of transition states on the pathway between reactants and products are not easily anticipated. This is not to say that they do not exhibit systematics as do "normal" molecules, but rather that we do not yet have sufficient experience to identify what systematics do exist, and more importantly how to capitalize on structural similarities. The problem is that transition states cannot even be detected let alone characterized experimentally, at least not directly. While measured activation energies relate to the energies of transition states above reactants, and while activation entropies and activation volumes as well as kinetic isotope effects imply some aspects of transition-state structure, no experiment can actually provide direct information about the detailed geometries of transition states.

Our only guide to the performance of theory in dealing with transition state geometries (aside from "reasonableness") is convergence with increasing level of calculation. This of course applies only to *ab initio* methods, which then need to serve as "benchmarks" for other calculations.

Only a few general remarks are appropriate at this time:

1. *Ab initio* Hartree-Fock models generally show a smooth progression in transition state geometries with increasing size of basis set. Bond lengths and skeletal bond angles typically show two to three times the sensitivity to changes in basis set than observed for equilibrium species.

2. Correlated (MP2) models generally but not always give "looser" transition states than Hartree-Fock schemes with the same basis set. There is insufficient experience to comment on convergence of MP2 models with respect to increasing size of basis set.

3. Semi-empirical models and to some extent minimal basis set *ab initio* models are unreliable in their description of transition state geometries. While transition states for many reactions, in particular for concerted processes, are similar to those from higher-level calculations, transition state geometries for other reactions can be quite different. Caution needs to be exercised in their use for this purpose.

4. There is insufficient experience to judge the performance of local density functional models for this purpose, although there are some indications that here transition structures may be too tightly bound. Caution needs to be exercised in their use.

4.3

Reaction Thermochemistry

Thermochemical comparisons may conveniently be placed into one of several categories depending on the extent to which bonds and non-bonded lone pairs are conserved (**Table 4-3**). This distinction is important as electronic structure models differ most in the way that they treat electron correlation, i.e., the coupling of motions of electrons, and correlation effects would be expected to be most important for electrons which are paired. Hartree-Fock models completely ignore correlation, while local density functional and MP2 models take partial explicit account of electron correlation. Semi-empirical models, which have been parameterized to reproduce experimental data, may be thought of as methods in which electron correlation is taken into account implicitly.

At one extreme, are processes in which not even the total number of electron pairs (bonds and non-bonded lone pairs) is conserved. Homolytic bond dissociation processes, e.g.,

$$H-F \rightarrow H\cdot + F\cdot \qquad \qquad \text{homolytic bond dissociation}$$

are an example. Comparisons of transition states and reactants (as required for the calculation of absolute activation energies) are also candidates for processes in which the total number of electron pairs is not conserved. This will be discussed in **Section 4.4**.

Less drastic are reactions in which the total number of electron pairs is maintained, but chemical bonds are converted to non-bonded lone pairs or vice versa. Heterolytic bond dissociation reactions, e.g.,

$$Na-F \rightarrow Na^+ + \ddot{F}^- \qquad \qquad \text{heterolytic bond dissociation}$$

and some structural isomerizations, e.g.,

$$H_2C=O \rightarrow HC\ddot{O}H \qquad \qquad \text{structural isomerism}$$

are examples. The most important examples may again involve comparisons between transition states and reactants (see **Section 4.4**).

Even more "gentle" are reactions in which both the number of bonds and the total number of non-bonded lone pairs are conserved. This type of reaction is very commonly encountered. Several examples are given below.

$$H_2C=CH_2 + 2H_2 \rightarrow 2CH_4 \qquad \qquad \text{hydrogenation}$$

$$\overline{CH_2CH_2CH_2} \rightarrow CH_3CH=CH_2 \qquad \qquad \text{structural isomerism}$$

$$2CH_2=CH_2 \rightarrow H_3C-CH_3 + HC\equiv CH \qquad \qquad \text{disproportionation}$$

Table 4-3: Performance of Theoretical Models for Description of Reaction Thermochemistry

type of process	examples	minimum level of calculation
no conservation of number of electron pairs	homolytic bond dissociation	correlated models excluding local density functional models; large basis sets
conservation of total number of electron pairs, but no conservation of number of bonds	heterolytic bond dissociation	Hartree-Fock and local density functional models; moderate to large basis sets with diffuse functions if anions are involved
conservation of total number of bonds and total number of non-bonded electron pairs, but no conservation of number of each kind of bond or number of each kind of non-bonded electron pair	hydrogenation, structural isomerism	Hartree-Fock and local density functional models; moderate to large basis sets
conservation of number of each kind of bond and number of each type of non-bonded electron pair (*isodesmic* reactions)	bond separation, remote substituent effects, regio- and stereochemical comparisons, conformation changes	Hartree-Fock models; small to moderate basis sets (local density functional models require moderate to large basis sets)

Most gentle are reactions in which the number of each kind of formal chemical bond (and each kind of non-bonded lone pair) are conserved. These are *isodesmic* ("equal bond") reactions. Examples include the processes below:

$$H_3C-C\equiv CH + CH_4 \rightarrow H_3C-CH_3 + HC\equiv CH \quad \text{bond separation}$$

$$(CH_3)_3NH^+ + NH_3 \rightarrow (CH_3)_3N + NH_4^+ \qquad \text{proton transfer}$$

In addition, all regio- and stereochemical comparisons are *isodesmic* reactions, as are conformation changes. Thus, *isodesmic* processes constitute a large class of reactions of considerable importance.

Many examples exist of all these types of reactions, and it is possible to offer some general remarks regarding the performance of various levels of calculation:

1. Correlated models (excluding local density functional models) are able to provide accurate descriptions of homolytic bond dissociation reactions. Large basis sets, with one or more sets of polarization functions and perhaps as well diffuse functions, are required. Hartree-Fock models and local density functional models, in particular, give unsatisfactory results. Hartree-Fock models underestimate the magnitudes of homolytic bond dissociation energies, while local density functional models overestimate their magnitudes. Present-generation semi-empirical models lead to unsatisfactory descriptions of the energetics of homolytic bond dissociation.

2. The energetics of reactions in which the total number of electron pairs are conserved, including heterolytic bond dissociation in which a bond is exchanged for a non-bonded lone pair, are generally well described using Hartree-Fock models. Moderate to large basis sets including polarization functions and, in the case of heterolytic bond dissociation where anions are produced, diffuse functions are required. Correlated models (including local density functional models) also perform well, although basis sets which are even larger than those needed for Hartree-Fock models may be required. Semi-empirical models give unsatisfactory results.

3. The energetics of *isodesmic* reactions are generally well described using both Hartree-Fock and correlated models (including local density functional models). Small to moderate basis sets usually give acceptable results for Hartree-Fock models, although larger basis sets may be required for use with correlated models (including local density functional models). Semi-empirical models give unsatisfactory results.

4. Hartree-Fock and correlated models (including local density functional models) with moderate basis sets perform reasonably well for conformation energy differences among stable rotomers or different ring conformations, as well as for single-bond rotation and inversion barriers. Semi-empirical models do not provide a reliable account.

Some additional comments follow from the observations above:

1. The fact that Hartree-Fock models provide reasonable descriptions of the energetics of processes in which the total number of electron pairs is conserved, together with the observation that Hartree-Fock reaction energies in these cases converge toward their limiting values much more rapidly than do correlated models (including local density functional models) with increasing size of basis set, suggests that they may be more suitable for this purpose than correlated models. That is to say, the description of the energetics of processes for which correlation effects largely cancel may be better accomplished with models which do not explicitly take electron correlation into account.

2. The poor performance of present-generation semi-empirical models for all types of thermochemical comparisons, even *isodesmic* reactions which are reasonably well described using minimal basis set Hartree-Fock models, may be due to the fact that they have been explicitly parameterized to minimize absolute errors in heats of formation rather than to minimize errors in reaction energies. Random errors in individual heats of formation are large enough (>7 kcal/mol in the case of AM1) such that the overall error in a given reaction will be unacceptable. Future generation semi-empirical models might be more successful for energetic comparisons were they to be parameterized to reproduce energies for specific classes of reactions.

The primary recommendation to follow from our comments (aside from needing to exercise caution in the use of semi-empirical models for energetic comparisons of any kind) is to make use of *isodesmic* reactions whenever possible. When this is not possible, try to write reactions in which the total number of chemical bonds is conserved.

4.4

Reaction Kinetics

Below we offer a few comments regarding the performance of different levels of calculation for both absolute and relative activation energies. Our remarks for relative activation energies in particular are based on very limited data and must be looked on as tentative.

1. Proper description of absolute activation energies requires correlated models (excluding local density functional models) with large underlying basis sets. Lower-level treatments are unreliable. Hartree-Fock models generally give absolute activation energies which are too large irrespective of basis set, while local density functional models usually underestimate absolute activation energies, sometimes significantly so. Semi-empirical models are unreliable in their account of absolute activation energies.

The assessment here parallels comments already made regarding the performance of the different models for homolytic bond dissociation energies (see **Section 4.3**).

2. Hartree-Fock models appear to provide a reasonable account of substituent effects on activation energies. Split-valence and polarization basis sets lead to acceptable results, but minimal basis set Hartree-Fock models and semi-empirircal models do not. There is insufficient experience to judge the performance of correlated models (including local density functional models) for this purpose, although there is no reason to believe that they too would not lead to acceptable results.

3. Hartree-Fock models, even with minimal basis sets, appear to give reasonable descriptions of the relative energies of transition states which differ only in regio- and/or stereochemistry. Semi-empirical models also seem to fare well in this regard. There is too little experience to assess the performance of correlated models (including local density functional models), although there is no reason to believe that they too will not lead to acceptable results.

5

Practical Strategies for Electronic Structure Calculations

Here we address a number of topics of importance to actually performing electronic structure calculations. Focus is on efficient determination of equilibrium and transition state geometries, and on the choice of geometry for use in thermochemical and kinetic comparisons. A more thorough discussion of practical strategies is available elsewhere.[5]

5.1

Finding Equilibrium Geometries

The energy of a molecule depends on its geometry, and even small changes in geometry can lead to large changes in total energy. Proper choice of molecular equilibrium geometry is therefore important in carrying out computational studies. What geometry should we use? Experimental structures would seem to be the best choice, given that they are available and are accurate. The trouble is that accurate gas-phase structure determinations are very tedious and have generally been restricted to very small molecules. While X-ray determinations on solid samples are more common, here one must be concerned about the role of the crystalline environment in altering geometry and particularly conformation. Ions present special problems; different counterions may lead to different geometries. Finally, only a few experimental structures exist for reactive "short-lived" molecules and for molecular complexes, among them hydrogen-bonded complexes.

Another approach might be to employ *idealized* or *standard* geometries. This is not unreasonable given the high degree of systematics exhibited by a large range of structures, in particular, structures of organic molecules. However, changes in properties among closely-related molecules may be very sensitive to subtle changes in geometry. For example, the likely reason that the dipole moment of trimethylamine is smaller than that in ammonia is the change in the local geometry around nitrogen. Were both molecules constrained to have the same geometry around nitrogen, then the relative magnitudes of the two dipole moments would probably not be reproduced. Another problem is that the structures of many of the most interesting molecules may differ greatly from the norm. All in all, assumed or standard geometries also do not offer a good solution.

In the final analysis, we usually will have little choice other than to obtain geometries directly from calculation. This is not as difficult as it might appear, at least if we have a reasonable idea where to start. Geometry optimization can be fully automated, and therefore requires no more human effort than a calculation utilizing an experimental or standard geometry.

Geometry optimization is an iterative process and several criteria must be satisfied before a geometry is accepted as optimized. First, successive geometry changes must not lower the total energy by more than a specified (small) value. Second, the energy gradient (first derivative of the energy with respect to geometrical distortions) at the optimized geometry must closely approach zero. Third, successive iterations must not change any of the geometrical parameters, i.e., bond lengths, angles, etc., by more than a specified (small) amount.

Efficient Optimization of Equilibrium Geometries

As pointed out in the previous chapter, semi-empirical and low-level Hartree-Fock models generally provide a reasonable account of molecular equilibrium geometries, as do higher-level Hartree-Fock and correlated models (including local density functional models). This being the case, it is nearly always advantageous to utilize a low-level model to provide a guess at equilibrium geometry and Hessian (matrix of second derivatives of the energy with respect to geometrical distortions). Typical results of the kinds of savings that are to be expected are provided in **Table 5-1**. Here the "low-level" model is AM1 and the "high-level" model is 3-21G. The reference (null) point is molecular mechanics using the SYBYL force field. We do not mean to imply that SYBYL (or molecular mechanics in general) is particularly bad for structure calculation, or that AM1 is particularly good, but only wish to point out the effect of starting geometry and Hessian on convergence.

Except for norbornane, all examples show significant reduction in the number of cycles required for optimization following the use of AM1 geometries and Hessians. Comparisons involving other Hartree-Fock, correlated or density functional methods would exhibit similar behavior, as would those where low-level Hartree-Fock models replace semi-empirical methods to provide guess geometry and Hessian. Given the low relative cost of semi-empirical schemes (full geometry optimization and Hessian calculation usually requiring less time than one optimization cycle of an *ab initio* calculation), it is clear that it is almost always advantageous to make use of them for this purpose.

SPARTAN facilitates such a strategy by allowing geometry optimization at one level of theory to be automatically preceded by optimization and/or Hessian calculation at a lower level. Thus, there is no additional "human time" required to take advantage of potentially large savings in computer time.

Table 5-1: **Number of Cycles Required for Geometry Optimization at the HF/3-21G Level as a Function of Starting Geometry and Hessian**

molecule	SYBYL geometry and SYBYL Hessian	AM1 geometry and AM1 Hessian
methylcyclopropyl ether	14	9
glycine	24	9
histidine	32	18
norbornane	6	5

5.2

Finding and Verifying Transition States

There are several reasons behind the common perception that finding transition states is inherently more difficult than finding equilibrium structures. Among the most important are:

1. The mathematical problem of locating a transition state is probably more difficult than that of finding a minimum. What is certainly true, is that techniques for locating transition states are less well developed than procedures for finding minima. After all, minimization is an important chore in many diverse fields of science and technology, whereas transition state location has few if any important applications outside of chemistry.

   Significant progress has been made in recent years in the development of methods and computer algorithms for locating transition states, and we can expect this progress to continue. In time, available techniques for transition state location should be of the same caliber as those for minima searching.

2. Potential energy surfaces in the vicinity of transition states are likely to be "flatter" than surfaces in the vicinity of local minima. This is entirely reasonable as transition states represent a delicate balance of bond breaking and bond forming, whereas overall bonding is maximized in equilibrium structures. The flatness of the potential energy surface complicates efforts to find transition states in two ways. First, it implies that the potential energy surface in the vicinity of a transition state will be less well described in terms of a simple quadratic function than the surface in the vicinity of a local minimum. This is important because all common optimization algorithms assume limiting quadratic behavior. Second, the geometries of transition states, even for closely related systems, or for the same system calculated at different levels of theory, might be expected to differ significantly. If true, this might limit the value of efforts to determine transition states for complex reactions based on structures for simpler processes and/or using lower levels of theory.

   There is a "silver lining". If the potential surface in the vicinity of the transition state is indeed flat, then large differences in transition-state geometries should have little effect on calculated activation energies. Semi-empirical or low-level *ab initio* models might provide suitable geometries for this purpose, even though their structural descriptions may differ significantly from those of higher-level models.

3. To the extent that transition states incorporate partially (or completely) broken bonds, it is to be anticipated that low-level theoretical treatments will not provide satisfactory energetics, and that models with large basis sets and which account for electron correlation will be required. While this is certainly the case with regard to calculated absolute activation energies, there is some evidence to suggest that even

low-level models properly account for activation energy differences among closely related reactions. Some discussion has already been provided in **Section 4.4**.

A related issue concerns the performance of parameterized semi-empirical models with regard to transition state geometries. Present-generation semi-empirical schemes, which have been parameterized using experimental data on stable molecules, would not be expected to perform as well in describing transition state structures as they would for equilibrium geometries, and as such perhaps cannot be trusted to provide reasonable guess geometries. This is not to say that they will not perform well, only that there is reason to be cautious.

4. We know relatively little about the geometries of transition states compared with our extensive knowledge about the geometries of stable molecules. Guessing transition state geometries based on prior experience is therefore much more difficult than guessing equilibrium geometries. This predicament, while obviously due in part to a complete lack of experimental structural data for transition states, also reflects a lag in the application of computational methods to the study of transition states (and reaction mechanisms in general). In one sense, our perception that finding transition states is difficult has contributed to the difficulty.

Guessing Transition State Geometries

There are several alternative ways to furnish a reasonable starting structure for a transition-state optimization:

1. Base the guess on the transition structure for a closely-related system which has previously been obtained at the same level of calculation, or alternatively on the transition structure for the same system obtained at a different (lower) level of theory. In dealing with "new reactions", where there is no prior experience, the best advice is to first gain experience with model systems and/or using low levels of calculation before finally proceeding with the systems or levels of calculation of actual interest.

2. Base the guess on an "average" of reactant and product geometries (**Linear Synchronous Transit** method). This is straightforward for unimolecular reactions, but requires that separated product and/or reactant molecules in bimolecular (and higher-order reactions) be replaced by a "weak complex" on the way to product (reactant). An example of this approach has earlier been provided in **Section 2.5**.

3. Base the guess on "chemical intuition", specifying critical bond lengths and angles in accord with preconceived notions of mechanism. Try not to impose symmetry on your structure guess, as this may limit its ability to alter in case your mechanistic assumption were to prove incorrect.

Verifying Transition States

Once a transition state has been located there are two tests which need to be performed in order to establish that it is both mathematically correct and indeed corresponds to the process of interest:

1. Verify that the Hessian yields one and only one imaginary frequency. This requires that a normal-mode analysis be carried out at the same level of calculation used to obtain the transition state; otherwise the results will be meaningless. The imaginary frequency will typically be in the range of 400-2000 cm^{-1}, i.e., quite similar to a real vibrational frequency. In case the molecule contains flexible rotors, e.g., methyl groups, or "floppy rings", the analysis may yield one or more additional imaginary frequencies with very small (<100 cm^{-1}) values. These typically correspond to couplings of low-energy motions and can usually be ignored; make certain that you establish to what motions these small imaginary frequencies actually correspond (see below) before doing so. Be wary of structures which yield only very small imaginary frequencies. This suggests a very low energy transition structure which more than likely will not correspond to the particular reaction of interest.

2. Verify that the coordinate corresponding to the imaginary frequency smoothly connects reactants and products. A simple way to do this is to "animate" the coordinate, that is, to "walk along" this coordinate without any additional optimization. This offers the advantage of not requiring any further calculation (beyond the normal-mode analysis already performed), but has the disadvantage that such an animation will not lead to the precise reactant or to the precise product. Experience suggests that this tactic is an inexpensive and effective way to eliminate transition states which do not connect the reactant with the desired product.

    Another approach is to actually "follow" the reaction from transition state to both the reactant and the product. A number of "coordinate following" schemes for doing this have been proposed, and these have collectively been termed **Intrinsic Reaction Coordinate** methods. Such schemes are not unique; while the reactant, product and transition state are well defined points of the overall potential energy surface, there are an infinite number of pathways linking them.

SPARTAN allows reaction coordinates to be followed by way of constrained optimization, and properties and graphics to be examined as a function of progression from reactant to product.

Efficient Optimization of Transition State Geometries

The same advice as already given with regard to determination of equilibrium geometries generally applies here. Whenever possible, precede optimizations at high levels of theory by optimizations using semi-empirical or small basis set Hartree-Fock models, and/or on simplified systems. Also recognize the value of a good Hessian in addition to a good geometry at the start of an optimization.

The effect of preceding 3-21G *ab initio* transition state optimizations by AM1 semi-empirical optimizations on the number of optimization cycles required is illustrated in **Table 5-2** for a few simple organic reactions. While it is unwise to generalize from such a small sample, it is clear that preliminary optimization using the AM1 model, followed by calculation of the Hessian at the AM1 level, effects significant reduction in total number of cycles required for optimization at the 3-21G level. We again want to emphasize that there is nothing special either about HF/3-21G as an optimization level or about AM1 as a means to supply guesses, and we use these levels here only for illustration. The overall conclusion should carry over to other *ab initio* schemes (including density functional and correlated methods), and to other semi-empirical and low-level *ab initio* models to supply the guess geometry and Hessian.

It does, however, need to be recognized that present-generation semi-empirical methods sometimes give very poor descriptions of the geometries of reaction transition states. They have not been explicitly parameterized for this purpose. Also, some low-level Hartree-Fock models and local density functional methods sometimes fail to produce reasonable transition state geometries. Caution needs to be exercised in the use of these methods for this purpose (see also **Section 4.2**).

Significant savings can also be achieved in some cases by preceding transition state optimization on the real system of interest by optimization on a closely-related model system. The idea here is that, even though transition state geometries vary somewhat from system to system, there may be a high degree of uniformity among closely-related systems.

The data in **Table 5-3** shows the effect of such a tactic on a few simple organic reactions for which models may easily be constructed. Here, calculations are at the semi-empirical AM1 level, although the conclusions should hold for other semi-empirical models, as well as for *ab initio* and correlated models, including local density functional models. While it is difficult to generalize from such a small data set, the value of preliminary semi-empirical calculations in reducing overall computation is clearly evident.

Table 5-2:   **Effect of Transition State Optimization and Hessian Calculation at the AM1 Level on the Number of Cycles Required for Transition State Optimization at the 3-21G Level**

reaction	optimization cycles required	
	no guess	AM1 guess
Claisen rearrangement of methyl, vinyl ether	100	20
Diels-Alder cycloaddition of cyclopentadiene and acrylonitrile	98	45
pyrolysis of ethyl formate	165	87

Table 5-3:   **Effect of Transition State Optimization on "Model Reactions" on the Number of Cycles Required for Transition State Optimization on "Real Reactions" at the AM1 Level**

real reaction	model reaction	optimization cycles required	
		no use of model	use of model
Diels-Alder cycloaddition of 5-chlorocyclopentadiene and acrylonitrile	Diels-Alder cycloaddition of cyclopentadiene and acrylonitrile	78	19
addition of singlet dichlorocarbene to cyclohexene	addition of singlet dichlorocarbene to ethylene	89	41
pyrolysis of cyclohexyl formate	pyrolysis of ethyl formate	291	191

5.3

Choice of Geometry for Thermochemical and Kinetic Comparisons

Thermochemical Comparisons

The favorable performance of low-level theoretical models (including semi-empirical models) in accounting for equilibrium geometry (see **Section 4.1**) suggests that they may in fact provide adequate structures for thermochemical comparisons carried out using more sophisticated models. This is actually a rather important practical issue, for optimization of equilibrium geometry can easily consume one or even two orders of magnitude more computation than energy (or property) calculation at a single geometry. It is legitimate to ask whether or not the substantial effort required to produce a "proper" optimized structure is effort well spent.

The examples in **Table 5-4** compare energies for a few simple chemical reactions obtained using both HF/6-31G* and MP2/6-31G* models*, each with three different sets of geometries: AM1, 3-21G and "exact" (6-31G* or MP2/6-31G*). Hydrogenation reactions, isomerization reactions and bond separation reactions are represented, to provide a reasonable sample of common processes. While it is not our purpose to assess the performance of the various theoretical models with regard to reaction thermochemistry, we have also provided experimental data. This allows comment on the magnitudes of errors brought about from the use of approximate geometries, relative to errors inherent to a particular theoretical model in describing a particular type of reaction.

Overall, errors incurred because of the use of approximate geometries for all three types of reactions considered are seldom greater than 1-2 kcal/mol, and generally much smaller than the differences between calculated and experimental reaction energies. The recommendation is quite clear: whenever possible, make use of either semi-empirical or low-level *ab initio* equilibrium geometries in constructing thermochemical comparisons based on higher-level models. While some caution is needed in dealing with systems where certain calculation levels are known to produce incorrect geometries, in general the errors resulting from the use of approximate geometries are very small.

Some properties exhibit greater sensitivity to exact choice of equilibrium geometries than total energies. For example, electric dipole moments for nitrogen and oxygen compounds (**Table 5-5**) alter somewhat with choice of geometry, although again the effects are not usually dramatic.

---

* Nomenclature: theory/basis set//theory/basis set. A single "slash" (/) separates designation of level of theory, e.g., HF for Hartree-Fock, MP2 for second-order Møller-Plesset correlated theory, from choice of basis set, e.g., 6-31G*. HF is optional and if missing implies Hartree-Fock theory. Semi-empirical models are designated by level of theory alone, e.g., AM1, without specification of basis set. A "double slash" separates the level of theory and basis set used for energy (property) calculation (on the left) from the level of theory and basis set used for geometry calculation, e.g., MP2/6-31G*//6-31G* implies an energy calculation at the MP2/6-31G* level using a (Hartree-Fock) 6-31G* geometry.

Table 5-4:   **Effect of Choice of Geometry on Energies of Reactions[a]**

	HF/6-31G*//			MP2/6-31G*//			
	AM1	3-21G	6-31G*	AM1	3-21G	MP2/6-31G*	expt.
hydrogenation reactions							
$CH_3CH_3 + H_2 \rightarrow 2CH_4$	-23	-22	-22	-18	-17	-16	-19
$CH_3NH_2 + H_2 \rightarrow CH_4 + NH_3$	-29	-27	-27	-25	-22	-23	-26
$CH_3OH + H_2 \rightarrow CH_4 + H_2O$	-29	-28	-27	-27	-25	-25	-30
$CH_3F + H_2 \rightarrow CH_4 + HF$	-18	-23	-23	-14	-22	-21	-29
$CH_2=CH_2 + 2H_2 \rightarrow 2CH_4$	-67	-66	-66	-60	-58	-58	-57
$CH_2=NH + 2H_2 \rightarrow CH_4 + NH_3$	-64	-60	-61	-56	-52	-52	-64
$CH_2=O + 2H_2 \rightarrow CH_4 + H_2O$	-58	-54	-54	-49	-46	-46	-59
$HC\equiv CH + 3H_2 \rightarrow 2CH_4$	-122	-121	-121	-109	-105	-104	-105
$HC\equiv N + 3H_2 \rightarrow CH_4 + NH_3$	-82	-77	-78	-64	-61	-60	-76
isomerization reactions							
formamide → nitrosomethane	63	65	64	67	61	61	62
acetonitrile → methylisocyanide	21	21	24	29	27	29	21
acetaldehyde → oxacyclopropane	30	33	31	27	28	27	26
acetic acid → methyl formate	20	14	13	14	14	14	18
ethanol → dimethyl ether	8	7	7	9	9	9	12
propene → cyclopropane	8	8	8	4	4	4	7
1,3-butadiene → bicyclo[1,1,0]butane	33	31	30	23	21	21	23
bond separation reactions							
$CH_3CH_2NH_2 + CH_4 \rightarrow CH_3CH_3 + CH_3NH_2$	3	3	3	5	4	4	3
$CH_3CH_2OH + CH_4 \rightarrow CH_3CH_3 + CH_3OH$	5	4	4	6	5	5	5
$CH_3CH=CH_2 + CH_4 \rightarrow CH_3CH_3 + CH_2=CH_2$	4	4	4	5	5	5	5
$CH_3CHO + CH_4 \rightarrow CH_3CH_3 + H_2CO$	10	10	10	11	11	11	11
$NH_2CHO + CH_4 \rightarrow CH_3NH_2 + H_2CO$	31	32	31	33	34	33	30
△ + 3CH_4 → 3CH_3CH_3	-24	-26	-26	-22	-24	-24	-22
△(NH) + 2CH_4 + NH_3 → CH_3CH_3 + 2CH_3NH_2	-21	-23	-22	-17	-18	-18	-17
△(O) + 2CH_4 + H_2O → CH_3CH_3 + 2CH_3OH	-19	-21	-22	-13	-14	-13	-14

a) reaction energies given in kcal/mol.

Table 5-5: **Effect of Choice of Geometry on Electric Dipole Moments[a]**

molecule	HF/6-31G*//			MP2/6-31G*//			expt.
	AM1	3-21G	6-31G*	AM1	3-21G	MP2/6-31G*	
Me_3N	0.64	0.73	0.75	0.54	0.64	0.74	0.61
Me_2NH	1.05	1.02	1.14	0.99	0.97	1.16	1.03
$EtNH_2$	1.44	1.26	1.49	1.39	1.22	1.54	1.22
$MeNH_2$	1.42	1.30	1.53	1.40	1.27	1.57	1.31
NH_3	1.80	1.55	1.92	1.80	1.55	1.97	1.47
$PhNH_2$	1.54	1.60	1.54	1.64	1.80	1.63	1.53
▷NH	1.83	2.01	1.94	1.72	1.87	1.90	1.90
⬡N	2.35	2.31	2.31	2.30	2.26	2.32	2.19
Me_2O	1.54	1.64	1.48	1.37	1.47	1.44	1.30
EtOH	1.78	1.80	1.74	1.66	1.67	1.68	1.69
MeOH	1.90	1.95	1.87	1.80	1.83	1.84	1.70
H_2O	2.25	2.18	2.20	2.20	2.14	2.20	1.85
▷O	2.35	2.64	2.28	2.01	2.24	2.11	1.89

a) dipole moments given in debyes.

"Exact" equilibrium structures **must be used** to determine vibrational frequencies and thermodynamic properties such as entropies obtained from these frequencies. The normal-mode analysis demands that the molecule be at a stationary point, and frequencies evaluated at non-equilibrium structures are meaningless!

Kinetic Comparisons

A similar question is whether or not it is necessary to utilize "exact" transition state geometries in determining activation energies, or whether instead geometries obtained from lower-level calculations will suffice. From a practical standpoint, success here is of even greater value than success in dealing with thermodynamic comparisons. Lack of experience with transition states makes guessing accurate starting geometries all the more difficult, and because of this optimization can be very time consuming. The issue comes down to balancing the possible magnitude of errors introduced as a result of using approximate geometries vs. the potentially enormous savings in computer time realized by employing low-level models to establish structure, in particular the structure of the transition state.

The data in **Table 5-6** illustrate for a small but diverse series of organic reactions the magnitudes of the errors in absolute activation energy introduced because of the use of approximate geometries. For each reaction, activation energies have been calculated using the MP2/6-31G* model with reactant and transition state geometries obtained at four different levels: semi-empirical AM1, Hartree-Fock 3-21G[(*)] and 6-31G* and correlated MP2/6-31G* models. Experimental absolute activation energies, where available, have also been provided.

The most obvious conclusion from the data is that both Hartree-Fock geometries and the MP2 geometry lead to very similar activation energies for all levels of calculation considered. Errors are typically in the range of 1-2 kcal/mol and seldom exceed 4 kcal/mol. The only serious exception occurs for 1,3-dipolar cycloaddition of formonitrile oxide and acetylene. Recognizing that the reactions presented here certainly do not adequately span the full range of possible processes, it is clear that even low-level Hartree-Fock models provide a credible account of the geometries both of reactants and transition states, and that the errors resulting from their use are consistently small. They should be used for this purpose wherever possible.

On the other hand, activation energies calculated assuming AM1 reactant and transition-state geometries sometimes differ significantly from those obtained from HF/3-21G[(*)], HF/6-31G* and MP2/6-31G* structures. The largest deviations are on the order of 10 kcal/mol, clearly an unacceptable error, although in about half the reactions considered the deviations are only on the order of 2-3 kcal/mol. Use of semi-empirical structures for the purpose of calculating absolute activation energies must, therefore, proceed only with some caution; only after acceptable results are achieved with model systems should they be trusted for related reactions.

A single example of the effect of choice of geometry on relative activation energies is provided in **Table 5-7**. This is for activation energies for Diels-Alder cycloadditions of cyclopentadiene (acting as a diene) with a variety of

Table 5-6:   **Effect of Choice of Geometry on Activation Energies at the MP2/6-31G* Level[a]**

reaction	geometry				
	AM1	3-21G	6-31G*	MP2/6-31G*	expt.
methylisocyanide → acetonitrile	43	44	42	43	38
ethyl formate → formic acid + ethylene	64	60	61	60	40,44
Cope rearrangement (1,5-hexadiene)	34	31	31	—	36
Claisen rearrangement (allyl,vinyl ether)	31	23	25	—	31
Diels Alder cycloaddition (cyclopentadiene + ethylene)	12	11	11	—	20
1,3 dipolar cycloaddition (formonitrile oxide + acetylene)	17	15	11	8	—
ene reaction (1-pentene)	60	61	61	60	—
CF_2 + ethylene → difluorocyclopropane	9	10	10	12	—
hydroboration (HBF_2 + ethylene)	33	28	27	20	—
oxacylohexenone → butadiene + carbon dioxide	43	43	45	—	—
cyclopentensulfone → butadiene + sulfur dioxide	16	26	27	—	—

a) activation energies given in kcal/mol.

Table 5-7:   **Effect of Choice of Geometry on Relative Activation Energies at the 6-31G* Level of Diels-Alder Cycloadditions of Cyclopentadiene with Electron-Deficient Dienophiles[a]**

dienophile	stereochemistry of adduct	geometry of reactant/transition state				
		AM1	STO-3G	3-21G	6-31G*	expt.
ethylene		4	4	4	4	8.5
cyanoethylene	endo	0	0	0	0	0
	exo	0	0	0	0	0
trans-1,2-dicyanoethylene		-4	-3	-3	-3	-2.6
cis-1,2-dicyanoethylene	endo	-4	-3	-3	-3	-3.8
	exo	-3	-3	-3	-3	-3.8
1,1-dicyanoethylene		-7	-7	-7	-8	-7.2
tricyanoethylene	endo	-9	-9	-9	-9	-9.2
	exo	-9	-9	-8	-9	-9.2
tetracyanoethylene		-12	-11	-11	-11	-11.2

a) activation energies given in kcal/mol.

cyano-substituted olefins (acting as dienophiles), relative to the *endo* addition with acrylonitrile as a standard. Energies have been evaluated at the HF/6-31G* level, with reactant and product geometries obtained from semi-empirical AM1 and Hartree-Fock STO-3G and 3-21G calculations, as well as from HF/6-31G* calculations ("exact" geometries). Experimental relative activation energies have been included for comparison.

The data clearly suggest that all models, including the AM1 model, provide suitable geometries for the calculation of relative activation energies in these systems. The largest error introduced is only 1 kcal/mol. The implication is that proper description of the relative activation energetics of reactions which differ from each other only in details of substitution can be obtained using approximate geometries. Unfortunately, the experimental data in this case do not distinguish between *endo* and *exo* pathways, so it is not possible to say whether the use of approximate geometries is adequate for the assignment of stereochemistry in these reactions.

Appendix A

Spartan

A.1

Architecture

SPARTAN is comprised of seven independent program modules: a graphical user interface, and **AB INITIO, DENSITY FUNCTIONAL, SEMI-EMPIRICAL, MECHANICS, PROPERTIES** and **GRAPHICS** modules. SPARTAN also interfaces with the Gaussian 94[9] and Mulliken[10] *ab initio*/density functional programs, and to Allinger's MM3[11] molecular mechanics program.

The figure below illustrates the interconnectivity of SPARTAN's modules.

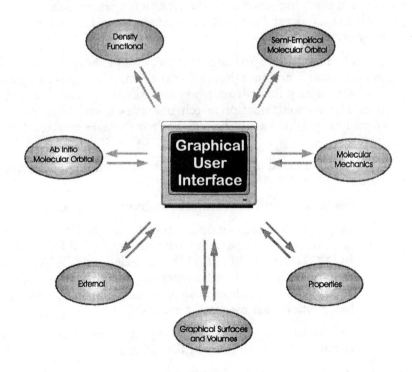

SPARTAN provides seamless interconnectivity between compute and graphical components. Without leaving the graphical interface, the user can construct a complex molecular structure, refine its geometry using molecular mechanics, specify a task and level of molecular mechanics or quantum chemical calculation, designate any graphical surfaces and/or any graphical volume, from which a surface or one of several different types of 2D slices can later be constructed for later display, specify properties of interest and then submit the calculation either to the local workstation, or to any server on the network. Once the calculation has completed, the user can view text and/or the graphical output. Requests for additional calculations and/or additional graphical and property output, may be made following completion. SPARTAN keeps track of what has already been done and will not repeat calculations unnecessarily. While one (or more) job is running, input for another job can be constructed, or output corresponding to yet another job viewed. In short, the graphical user interface is a window into SPARTAN, allowing convenient access to its many features.

SPARTAN's architecture makes a clear separation between **tasks** and **methods**. Tasks indicate what is to be done, e.g., perform a geometry optimization or search conformation space, while methods dictate how the tasks are to be done, e.g., use MM3 molecular mechanics or the PM3 semi-empirical model. Both tasks and methods are specified in the graphical user interface, while the actual calculations are performed in the outlying computation modules. In principle, any task can be handled by any method and, while practical considerations may prove limiting in some cases, e.g., full conformation searches using high level *ab initio* techniques are probably impractical, the notion that tasks and methods are independent is fundamental in the design of SPARTAN.

SPARTAN also allows operations to be set up and performed on collections of molecules with no more (human) effort than required for single molecules. Collections may result from previous tasks, e.g., a collection of structures results from a conformation search, or may be user defined, e.g., a collection of molecules which are to be searched for a similar structural or electronic feature.

SPARTAN's graphical user interface provides a number of functions, among them:

1. the construction and editing of molecular structures,

2. the preparation of input designating the quantum chemical or molecular mechanics calculation to be performed by the **AB INITIO, DENSITY FUNCTIONAL, SEMI-EMPIRICAL** or **MECHANICS** modules; the preparation of input for external programs, presently the Gaussian 94[9] and Mulliken[10] *ab initio*/density functional programs, and Allinger's MM3[11] molecular mechanics program,

3. the preparation of input for designating molecular properties to be calculated using the **PROPERTIES** module,

4. the preparation of input for designating graphical surfaces and/or volumes to be constructed by the **GRAPHICS** module for later display in the interface,

5. the display of text output resulting from molecular mechanics and quantum chemical calculations,

6. the display and manipulation of structures resulting from molecular mechanics and quantum chemical calculations,

7. the display of isosurfaces from surface data, and construction and display of 2D slices and isosurfaces from volume data for single molecules or as sums and differences of volume data between molecules,

8. the animation of normal modes of vibration, or motion along other geometrical coordinates, e.g., torsional motion; both geometrical structures and any graphical displays may be animated,

9. setting up of coordinate sequences used, for example, to simulate progression from reactant to product through a transition state,

10. setting up collections of molecules, either as a result of previously completed tasks, or user defined,

11. similarly searching among molecules, based either on geometrical structure or on any previously defined volume of data,

12. the "export" and "import" of structures and other data to and from external programs,

13. the printing of graphical displays, and

14. the display of the electric dipole moment vector.

SPARTAN's **AB INITIO, DENSITY FUNCTIONAL**, and **SEMI-EMPIRICAL** modules each serve four primary functions:

1. the calculation of the energy (heat of formation in the case of the **SEMI-EMPIRICAL** module) and wavefunction corresponding to a single geometry,

2. the calculation of equilibrium and transition state geometries,

3. the calculation of the Hessian (matrix of second derivatives) required for evaluation of normal mode vibrational frequencies and calculation of thermodynamic properties (the actual frequency and thermodynamic calculations are performed by the **PROPERTIES** module), and

4. searching of conformation space.

SPARTAN's **MECHANICS** module serves three primary functions:

1. the calculation of the strain energy corresponding to a single geometry,

2. the calculation of equilibrium geometry, and

3. searching of conformation space.

SPARTAN's **PROPERTIES** module serves five primary functions:

1. preparation of text output,

2. population analyses via the Mulliken[12] and/or natural bond orbital[13] procedures, and charge calculations based on fitting to molecular electrostatic potentials,[14]

3. calculation of normal modes of vibration,

4. evaluation of thermodynamic properties, and

5. calculation of the electric dipole moment.

SPARTAN's **GRAPHICS** module is responsible for the actual calculation (but not the display) of volumes and surfaces and properties encoded onto those surfaces, based either on *ab initio*, density functional or semi-empirical wavefunctions. These include total electron and spin densities, and molecular electrostatic potentials and polarization potentials, as well as the molecular orbitals.

A.2

Capabilities and Limitations

AB INITIO Module

SPARTAN's **AB INITIO** module performs Hartree-Fock and MP2 calculations[15] for both closed-shell and open-shell (unrestricted) systems. Internally stored basis sets include STO-3G, 3-21G, 6-31G and 6-311G, with extensions for polarization functions and/or diffuse basis functions[16]. Input of an arbitrary basis set, comprising s-, p- and d-type Gaussians, is permitted.

Preset limits within SPARTAN's **AB INITIO** module are enumerated below.

maximum number of atoms	100
maximum number of basis functions	600

DENSITY FUNCTIONAL Module

SPARTAN's **DENSITY FUNCTIONAL** module presently performs local density functional calculations. Three sets of numerical basis are available for the first 54 elements of the Periodic Table. These correspond approximately to 6-31G, 6-31G* and 6-31G** Gaussian basis sets.

Preset limits within SPARTAN's **DENSITY FUNCTIONAL** module are enumerated below.

maximum number of atoms	100
maximum number of basis functions	600

SEMI-EMPIRICAL Module

SPARTAN's **SEMI-EMPIRICAL** module performs MNDO[17], MNDO/d[18], AM1[19] and PM3[20] calculations. Also available is a new semi-empirical model for transition metals called PM3 (tm)[21]. This describes each transition metal in terms of explicit d-type functions as well as s- and p-type valence atomic orbitals, and as such is distinct from previous semi-empirical parameterizations for (right-hand) transition metals. PM3 (tm) is intended to be used in conjunction with PM3 for non-transition metals.

Corrections for aqueous solvation using the SM1, SM1a and SM2 models developed by Cramer and Truhlar may be obtained using AM1 wavefunctions[22]. The Cramer/Truhlar SM3 model for water may be used with the PM3 method[22]. Also incorporated in *SPARTAN* are solvation models for both water and hexadecane developed by Dixon, Leonard and Hehre[23]. These have been specifically parameterized for the MNDO, AM1 and PM3 methods (six parameterizations in total).

Preset limits for *SPARTAN*'s **SEMI-EMPIRICAL** module are enumerated below.

maximum number of atoms (any type)	200

MECHANICS Module

SPARTAN's **MECHANICS** module supports MM2[24], MM3[24] and SYBYL[25] force fields. At the present time, corrections for conjugated π systems have not been included in *SPARTAN*'s implementation of the MM2 and MM3 force fields (corrections to MM3 are supported using *SPARTAN*'s interface to Allinger's MM3 program).

There are no preset limits for *SPARTAN*'s **MECHANICS** module.

Appendix B

Operation of *Spartan*

Here we review briefly some of the operating features of *Spartan*. Full details are available in the *Spartan* **User's Guide**.[6]

B.1

Starting and Quitting *Spartan*

To intiate *Spartan*, bring up a "window" and type **spartan**. In a few seconds another "window" will appear. Move it into the desired screen location, and *click* with any mouse button. This results in a clear screen, except for a menu bar across the top.

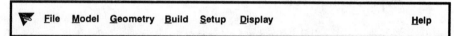

The display which appears may be moved about the screen by positioning the cursor in the area immediately above the menu bar, depressing either the left or the middle mouse button, and moving the mouse. It may be scaled by *grabbing* one of the corners, depressing either the left or the middle mouse button, and moving the mouse.

To exit *Spartan*, select **Quit** from the **File** menu (see **B.2** below).

B.2

Pull-Down Menus

Most of *Spartan*'s program functions are accessed using *pull-down* menus under the headings in the menu bar. Position the cursor on top of the appropriate item in the menu bar, and *click* with the left mouse button. The menu will appear below, e.g., *clicking* on **Build**.

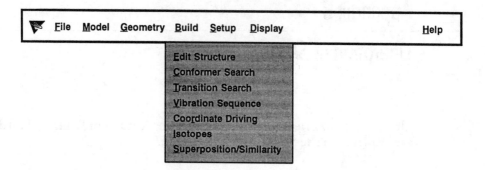

Click on the appropriate item in the menu to select it. Alternatively, slide the cursor while holding down on the left mouse button, first to the desired menu, and then to the desired item under the menu, and only then releasing the button. The latter mode corresponds to the operation of a MAC.

B.3

Keystroke Equivalents

Functions available under *SPARTAN*'s menus may also be accessed from the keyboard. Keystroke equivalents for each of the menus are designated by the underlined letter in each of the menu names, i.e,. **File, Model, Geometry, Build, Setup, Display,** and **Help,** and the menus are accessed by simultaneously depressing the **Alt** key, and the corresponding letter, e.g., **Alt B** accesses the **Build** menu. **Alt W** accesses the Wavefunction **Logo** menu. Each of the items in the individual menus also incorporates a letter which has been underlined, e.g., in the case of the **Build** menu,

Edit Structure
Conformer Search
Transition Search
Vibration Sequence
Coordinate Driving
Isotopes
Superposition/Similarity

The item is accessed by depressing this key. For example, to enter the **Conformer Search** dialog, first simultaneously depress the **Alt** and **B** keys to access the **Build** menu, and then press the **C** key.

B.4

Use of the Mouse

Enumerated below are functions associated with the three-button mouse:

keyboard	mouse buttons		
	left	middle	right
—	picking	X/Y rotate	X/Y translate
Shift key	—	Z rotate	scale
space bar	—	rotate bond	—
Control key	—	global (X/Y) rotate (Z rotate with **Shift key**)	global (X/Y) translate (Z translate with **Shift key**)

The left mouse button is for picking (of graphical objects and/or menu items), the middle button for rotation of objects, and the right button for translation of objects. Rotation and translation functions may be modified by holding down specific keys in addition to the appropriate mouse buttons. Together with the **Shift key**, the middle mouse button results in rotation of the molecule about the Z direction (perpendicular to the screen), while the right mouse button results in scaling. The **space bar**, in addition to the middle mouse button, allows for rotation about a selected bond (in SPARTAN's molecule builders).

The **Control key** ("Ctrl"), in conjunction with the middle or right mouse buttons and (optionally) the **Shift key**, signifies that rotations or translations are to be carried out on all molecules presently displayed rather than only on the selected molecule.

B.5

Molecule Selection

While two or more molecules may be simultaneously displayed in SPARTAN's main window, only one molecule may be selected. The selected molecule has access to all menu capabilities (molecule building, job setup and submission, and text and graphical display, and manipulation), while non-selected molecules may only be displayed as static images. Selection of one of the several sets of images currently on screen occurs by moving the cursor anywhere onto the image, i.e., onto the structure representation or onto any graphical surface, and *clicking* with the left mouse button. This results in de-selection of the prior set of images.

B.6

Collections of Molecules

Selection need not refer to an individual molecule but rather to a collection of molecules. Accompanying the collection is a dialog box (the **Lists** dialog),

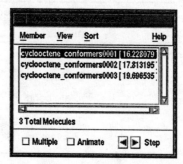

which contains a series of menus to allow list editing (**Member**), control the information displayed (**View**), and sort the members (**Sort**). The main body is given over to a box which lists the individual molecules in the collection. The **Multiple** switch toggles between two different display modes; with **Multiple** "off" only one member of the list may be displayed, while with **Multiple** "on" several members may be simultaneously displayed.

Molecule selection with **Multiple** "off" follows by *clicking* on the name which appears in the scroll box. It is then highlighted (displayed in *reverse video*) and the corresponding structure model displayed on screen. The list entry corresponding to the previously selected entry is de-highlighted, and its structural model removed from the screen. The **Step** keys may be used to "walk through" the list, and the **Animate** key may be used to "walk through" the list in rapid succession.

Selection with **Multiple** "on" occurs by *clicking* on an entry in the list, after which it is highlighted, and the structure model displayed on screen. Additional entries may be selected for simultaneous display by *clicking* on them. De-selection occurs by *clicking* on a list entry which is already selected (highlighted), after which it is de-highlighted, and its structural model removed. The exception is that at least one member of the list needs to be displayed, and *clicking* on the last remaining highlighted entry will have no effect. The **Step** and **Animate** keys are not accessible with **Multiple** "on".

B.7

Changing Colors and Setting Preferences

The Wavefunction **Logo** menu 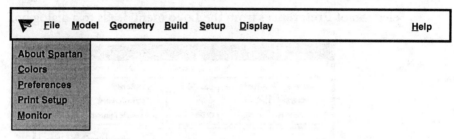 provides, among other things, for changing background and graphical object colors as well as for setting program defaults.

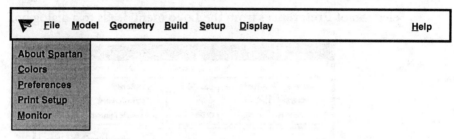

Selection of **Colors** from the ◥ menu leads to an on-screen message,

> Colors: Select object. (Enter "." to abort)

which directs the user to pick a graphical object on the screen. This accomplished, a second on-screen message,

> Colors: Select color. (Enter "." to abort)

and a dialog box appears.

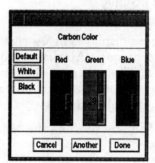

The graphical object selected is identified at the top of the dialog, e.g., "Carbon Color". On the left are three buttons, *clicking* on which accesses predefined colors: **Default** (which depends on the graphical object selected), **White** and **Black**. Object color may also be set by adjusting the **Red**, **Green** and **Blue** *slider bars* in the center of the dialog. Finally, the color of the previously selected graphical object may be matched to the color of any monochrome object on screen (atoms, bonds and surfaces without encoded properties), or screen background colors by moving the cursor anywhere on screen (except inside a dialog box) and *clicking*.

Color selection options may be performed in any order, and as many times as desired. Once completed, *clicking* on **Another** sets the color and displays a message requesting selection of another graphical object. *Clicking* on **Done** sets the color and exits the dialog. *Clicking* on **Cancel** exits the dialog without instituting the last (since the last *click* on **Another**) color change.

Selection of **Preferences** from the **Logo** menu leads to a dialog box.

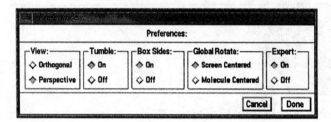

View controls whether objects are displayed as orthogonal or perspective representations, **Tumble** controls automatic tumbling of images on screen, **Box Sides** indicates the type of box (wire frame or transparent solid), **Global Rotate** sets the center for simultaneous rotation of multiple objects on screen and **Expert** controls access to SPARTAN's "expert" dialogs.

B.8

3-Dimensional Displays

Any graphical object displayed in SPARTAN may be rendered in stereo using color filtration techniques. All that is required is for the viewer to wear "red/blue" glasses. While there is obviously loss of color in the resulting image, this simple technique works almost as well as polarization techniques which require additional hardware.

Stereo is turned "on" by pressing the "**3**" key, and turned "off" by again pressing the "**3**" key.

B.9

Iconizing SPARTAN

At any time during execution, SPARTAN may be *iconized* by clicking on the "⬛" box at the top right of the screen. This returns the user to the window from which SPARTAN was initiated, and places a small icon on the screen.

Clicking on this icon returns the user to SPARTAN.

Appendix C

Directory of Spartan's Main Menus

Logo

About Spartan	Provides copyright information and addresses
Colors	Selects screen and model colors
Preferences	Selects modes for model display
Print Setup	Sets up printers
Monitor	Monitors, suspends and resumes and kills jobs

File

New	Enters the builders
Open	Opens a molecule
Close	Closes a molecule
Save As	Saves a molecule
Merge As	Merges all molecules on screen into a single "molecule"
Group As	Groups all molecules on screen into a list of molecules
Delete	Deletes an existing molecule
Import	Imports a molecule from an external program
Export	Exports a molecule to an external program
Print	Prints on-screen display
Quit	Exits Spartan

Model

Wire	Displays structure as wire-frame model
Ball and Wire	Displays structure as ball-and-wire model
Tube	Displays structure as tube model
Ball and Spoke	Displays structure as ball-and-spoke model
Space Filling	Displays structure as space-filling (CPK) model
Hide	Removes present model from screen
Hide/Show Hydrogens	Removes (Hide Hydrogens) and displays (Show Hydrogens) hydrogens in the structure model
Show/Hide Labels	Displays (Show Labels) and removes (Hide Labels) atom labels

Geometry

Distance	Displays distance between two atoms
Angle	Displays angle involving three atoms or two bonds
Dihedral	Displays dihedral angle involving four atoms or three bonds
Distance to Plane	Displays distance between an atom and a plane
Angle with Plane	Displays angle involving two atoms or a bond and a plane
Define Point	Defines a point as a geometric mean of a set of atoms
Define Plane	Defines a plane made by three atoms
Show/Hide Box	Displays (Show Box) and hides (Hide Box) a frame around a molecule
Report Symmetry	Reports symmetry point group

Build

Edit Structure	Re-enters the builders
Conformer Search	Specifies method and limits of conformation search
Transition Search	Generates a guess at transition state geometry based on reactant and product geometries
Vibration Sequence	Generates sequence of structures along normal coordinate
Coordinate Driving	Sets up a sequence of geometry constraints
Isotopes	Changes default atomic masses
Superposition/Similarity	Superimposes molecules based on similarity in structure or some graphical indicator

Setup

Lists	Controls list-processing (in Conformer Search, Coordinate Driving and Superposition/Similarity)
Ab Initio	Sets up *ab initio* calculations
Density Functional	Sets up density functional calculations
Semi-Empirical	Sets up semi-empirical calculations
Mechanics	Sets up molecular mechanics calculations
External	
Gaussian94	Sets up jobs for Gaussian 94 electronic structure program
Mulliken	Sets up jobs for Mulliken electronic structure program
MM3	Sets up jobs for Allinger's MM3 molecular mechanics program
Properties	Sets up calculation of properties
Surfaces	Sets up generation of graphical surfaces
Volumes	Sets up generation of volumes of graphical data
Submit	Submits a job for execution

Display

Output	Displays text output
Properties	
Energy	Reports total energy
Dipole	Reports dipole moment and displays dipole moment vector
Charge	Reports atomic charges
Surfaces	Reports the value of a property encoded onto a surface or 2D slice; displays surface area and volume
Surfaces	Displays an isosurface with or without an encoded property created from surface data
Slices	
Create	Creates a 2D slice from volume data
Edit	Edits a 2D slice
Delete	Deletes a 2D slice
Isosurfaces	
Create	Creates an isosurface with or without an encoded property from volume data
Edit	Edits an isosurface
Delete	Deletes an isosurface
Vibrations	Animates normal-mode vibrations

Help

Use of the Mouse	Describes the use of the mouse
Keystroke Equivalents	Lists keystroke equivalents to *Spartan's* menus
Ab Initio Options	Lists options available in the *ab initio* module
Density Functional Options	Lists options available in the density functional module
Semi-Empirical Options	Lists options available in the semi-empirical module
Mechanics Options	Lists options available in mechanics module
Properties Options	Lists options available in properties module
Graphics Options	Lists options available in graphics module
Conformer Options	Lists options available for conformer searching
Superposition Options	Lists options available for superposition/similarity
Available Elements	Lists available elements for *ab initio*, density functional, semi-empirical and molecular mechanics calculations

Collections of molecules involve an additional (**Lists**) dialog with the following menus.

Member

Extract As	Extracts one or more members from a list
Delete	Deletes a list member
Delete to End	Deletes list members from the selected member to the last member

---

View

Hide/Show List	Displays (Show List) and hides (Hide List) list of molecules
Show/Hide Energy	Displays (Show Energy) and hides (Hide Energy) energy
Decouple/Couple Coordinates	Decouples (Decouple Coordinates) and couples (Couple Coordinates) coordinates

---

Sort

by Energy	Sorts members of the list by energy
by Name	Sorts members of the list by name

Help

Overview	Describes the operation of the Lists dialog

References

1. W.J. Hehre, L.D. Burke, A.J. Shusterman, *A SPARTAN Tutorial*, Version 3.0, Wavefunction, Irvine, 1993; W.J. Hehre, W.W. Huang, L.D. Burke, A.J. Shusterman, *A SPARTAN Tutorial*, Version 4.0, Wavefunction, Irvine, 1995.

2. W.J. Hehre, L.D. Burke, A.J. Shusterman and W.J. Pietro, *Experiments in Computational Organic Chemistry*, Wavefunction, Irvine, 1993.

3. W.J. Hehre, L. Radom, P.v.R. Schleyer and J.A. Pople, *Ab Initio Molecular Orbital Theory*, Wiley, New York, 1986.

4. W.J. Hehre, *Critical Assessment of Modern Electronic Structure Methods*, Wavefunction, Irvine, 1995.

5. W.J. Hehre, *Practical Strategies for Electronic Structure Calculations*, Wavefunction, Irvine, 1995.

6. *SPARTAN Users Guide, Version 4.0*, Wavefunction, Irvine, 1995.

7. R.G. Parr and W. Yang, *Density Functional Theory of Atoms and Molecules*, Oxford University Press, New York, 1989.

8. (a) T. Clark, *A Handbook of Computational Chemistry*, Wiley, New York, 1986; (b) J.J.R. Stewart, J. Computer Aided Molecular Design, 4, 1 (1990).

9. Gaussian 94, available from: Gaussian, Inc., Carnegie Office Park, Bldg. 6, Pittsburgh, PA 15106.

10. Mulliken, available from: CAChe Scientific, P.O. Box 500, Beaverton, OR 97077.

11. MM3, available from: Tripos Associates, 1699 S. Hanley Road, St. Louis, MO 63144.

12. R.S. Mulliken, J. Chem. Phys., 23, 1833, 1841, 2338, 2343 (1955).

13. (a) J.P. Foster and F. Weinhold, J. Am. Chem. Soc., 102, 7211 (1980); (b) A.E. Reed and F. Weinhold, J. Chem. Phys. 708, 4066 (1983); (c) A.E. Reed, R.B. Weinstock and F. Weinhold, *ibid.*, 83, 735 (1985); (d) J.E. Carpenter and F. Weinhold, J. Mol. Struct. (Theochem.), 169, 41 (1988).

14. (a) L.E. Chirlian and M.M. Francl, J. Computational Chem., 8, 894 (1987); (b) C.M. Breneman and K.B. Wiberg, *ibid.*, 11, 361 (1990).

15. Additional electron correlation techniques including higher-order Møller Plesset models and CI models are available from Gaussian 94[9], which may be accessed transparently from *SPARTAN*'s graphical user interface.

16. For references to the basis sets supported in *SPARTAN*, see ref. 3, Chapter 4.

17. M.J.S. Dewar and W.J. Thiel, J. Am. Chem. Soc., 99, 4899 (1977).

18. (a) W. Thiel and A. Voityuk, Theor. Chem. Acta., 81, 391 (1992); (b) W. Thiel and A. Voityuk, Int. J. Quantum Chem., 44, 807 (1992).

19. M.J.S. Dewar, E.G. Zoebisch, E.F. Healy and J.J.P. Stewart, J. Am. Chem. Soc., 107, 3902 (1985).

20. J.J.P. Stewart, J. Computational Chem., 10, 209 (1989).

21. J. Yu and W.J. Hehre, J. Computational Chem. submitted.

22. (a) C.J. Cramer and D.G. Truhlar, J. Am. Chem. Soc., **113**, 8305 (1991); (b) C.J. Cramer and D.G. Truhlar, Science, **256**, 213 (1992); (c) C.J. Cramer and D.G. Truhlar, J. Comp. Aid. Mol. Des., **6**, 69 (1992).

23. (a) R.W. Dixon, J.M. Leonard and W.J. Hehre, Israel J. Chem., **33**, 427 (1993); (b) R.W. Dixon, J.M. Leonard and W.J. Hehre, J. Am. Chem. Soc., submitted.

24. Review: J.P. Bowen and N.L. Allinger, Revs. in Computational Chem., **2**, 81 (1991).

25. M. Clark, R.D. Cramer III and N. van Opdensch, J. Computational Chem., **10**, 982 (1989).

Index

(Entries in **bold type** correspond to functions under *SPARTAN*'s menus. The reference is to the first-time usage.)